William Ashbrook Kellerman

Spring flora of Ohio consisting of descriptions of the early native flowering plants

William Ashbrook Kellerman

Spring flora of Ohio consisting of descriptions of the early native flowering plants

ISBN/EAN: 9783337268497

Printed in Europe, USA, Canada, Australia, Japan

Cover: Foto ©berggeist007 / pixelio.de

More available books at **www.hansebooks.com**

SPRING

FLORA OF OHIO

CONSISTING OF

DESCRIPTIONS OF THE EARLY NATIVE FLOWERING PLANTS, WITH
KEYS FOR THEIR IDENTIFICATION; ALSO A KEY FOR THE
IDENTIFICATION OF THE TREES AND SHRUBS OF
THE STATE BY THEIR LEAVES AND FRUITS

BY

W. A. KELLERMAN, Ph. D.
Professor of Botany in the Ohio State University

PUBLISHED BY THE AUTHOR

COLUMBUS, OHIO
J. L. TRAUGER, PRINTER AND PUBLISHER
1895

SEQUENCE OF ORDERS.

I.	Coniferæ.		XLVI.	Anacardiaceæ.
II.	Araceæ.		XLVII.	Aquifoliaceæ.
III.	Commelinaceæ.		XLVIII.	Celastraceæ.
IV.	Liliaceæ.		XLIX.	Staphyleaceæ.
V.	Smilaceæ.		L.	Aceraceæ.
VI.	Amaryllidaceæ.		LI.	Hippocastanaceæ.
VII.	Dioscoreaceæ.		LII.	Rhamnaceæ.
VIII.	Iridaceæ.		LIII.	Vitaceæ.
IX.	Orchidaceæ.		LIV.	Tiliaceæ.
X.	Juglandaceæ.		LV.	Malvaceæ.
XI.	Myricaceæ.		LVI.	Hypericaceæ.
XII.	Salicaceæ.		LVII.	Cistaceæ.
XIII.	Betulaceæ.		LVIII.	Violaceæ.
XIV.	Fagaceæ.		LIX.	Passifloraceæ.
XV.	Ulmaceæ.		LX.	Thymelæaceæ.
XVI.	Moraceæ.		LXI.	Elæagnaceæ.
XVII.	Loranthaceæ.		LXII.	Lythraceæ.
XVIII.	Santalaceæ.		LXIII.	Araliaceæ.
XIX.	Aristolochiaceæ.		LXIV.	Umbelliferæ.
XX.	Polygonaceæ.		LXV.	Cornaceæ.
XXI.	Aiziodaceæ.		LXVI.	Ericaceæ.
XXII.	Portulacaceæ.		LXVII.	Primulaceæ.
XXIII.	Caryophyllaceæ.		LXVIII.	Ebenaceæ.
XXIV.	Nymphæaceæ.		LXIX.	Oleaceæ
XXV.	Magnoliaceæ.		LXX.	Gentianaceæ.
XXVI.	Anonaceæ.		LXXI.	Apocynaceæ.
XXVII.	Ranunculaceæ.		LXXII.	Polemoniaceæ.
XXVIII.	Berberidaceæ.		LXXIII.	Convolvulaceæ.
XXIX.	Calycanthaceæ.		LXXIV.	Hydrophyllaceæ.
XXX.	Lauraceæ.		LXXV.	Boraginaceæ.
XXXI.	Papaveraceæ.		LXXVI.	Labiatæ.
XXXII.	Cruciferæ.		LXXVII.	Solanaceæ.
XXXIII.	Saxifragaceæ.		LXXVIII.	Scrophulariaceæ.
XXXIV.	Hamamelidaceæ.		LXXIX.	Lentibulariaceæ.
XXXV.	Platanaceæ.		LXXX.	Orobanchaceæ.
XXXVI.	Rosaceæ.		LXXXI.	Bignoniaceæ.
XXXVII.	Leguminosæ.		LXXXII.	Pedaliaceæ.
XXXVIII.	Geraniaceæ.		LXXXIII.	Acanthaceæ.
XXXIX.	Oxalidaceæ.		LXXXIV.	Plantaginaceæ.
XL.	Rutaceæ.		LXXXV.	Rubiaceæ.
XLI.	Simarubaceæ.		LXXXVI.	Caprifoliaceæ.
XLII.	Polygalaceæ.		LXXXVII.	Valerianaceæ.
XLIII.	Euphorbiaceæ.		LXXXVIII.	Campanulaceæ.
XLIV.	Callitrichaceæ.		LXXXIX.	Compositæ.
XLV.	Limnanthaceæ.		XC.	Cichoriaceæ.

INDEX TO ORDERS AND GENERA.

	PAGE		PAGE		PAGE
Acanthaceæ	110	Aruncus	63	Catalpa	109
Acer	77	Asarum	37	Caulophyllum	49
Aceraceæ	77	Asimina	42	Ceanothus	79
Achillea	120	Asparagus	21	Celastraceæ	76
Aconitum	44	Azalea	91	Celastrus	77
Actæa	44	Barbarea	53	Celtis	36
Adopogon	121	Batrachium	47	Cephalanthus	112
Aesculus	79	Baptisia	69	Cerastium	39
Ailanthus	73	Belamcanda	25	Cercis	69
Aizoidaceæ	37	Benzoin	30	Chærophyllum	88
Allium	20	Berberidaceæ	48	Chamædaphne	92
Alnus	32	Berberis	49	Chamælirium	20
Alsine	39	Betula	32	Chelidonium	50
Amelanchier	62	Betulaceæ	31	Chiogenes	93
Amaryllidaceæ	24	Bicuculla	51	Chionanthus	97
Ampelopsis	80	Bignonia	109	Chrysanthemum	120
Anacardiaceæ	75	Bignoniaceæ	108	Chrysosplenium	59
Anagallis	95	Boraginaceæ	100	Cichoriaceæ	121
Andromeda	92	Brunellâ	104	Cimicifuga	44
Anemone	45	Buda	40	Cistaceæ	82
Anemonella	45	Buettneria	49	Claytonia	38
Angelica	87	Bursa	55	Clematis	46
Anonaceæ	42	Callitrichaceæ	74	Collinsia	105
Antennaria	119	Callitriche	75	Comandra	37
Anthemis	120	Caltha	43	Comarum	66
Anychia	41	Calycanthaceæ	49	Commelinaceæ	19
Aphyllon	108	Calycanthus	49	Compositæ	118
Aplectrum	27	Camassia	21	Comptonia	29
Apocynaceæ	97	Campanula	118	Conopholis	108
Apocynum	98	Campanulaceæ	118	Convallaria	23
Aquifoliaceæ	76	Capnoides	51	Convolvulaceæ	99
Aquilegia	44	Caprifoliaceæ	113	Convolvulus	99
Arabis	56	Capsella	55	Coptis	43
Araceæ	18	Cardamine	54	Corallorhiza	27
Aralia	86	Carpinus	31	Cornaceæ	89
Araliaceæ	85	Carya	28	Cornus	89
Arctostaphylos	93	Caryophyllaceæ	38	Corydalis	51
Arenaria	40	Cassandra	92	Corylus	31
Arisæma	19	Castalia	41	Cratægus	65
Aristolochiaceæ	37	Castanea	33	Cruciferæ	51
Aronia	62	Castilleja	107	Cynoglossum	101

to only so far generally as is necessary to contrast and identify our species. The constant aim has been to select the most obvious and striking available characters which the beginner can in no case fail to comprehend readily. The names have been divided into syllables and the accented syllables marked, thus making the correct pronunciation of the supposed difficult names an easy and simple matter.

The scientific names of the plants are here given as they have recently been agreed upon by North American botanists. The names used have been determined according to the principle of priority. This principle is, briefly stated, that the oldest, i. e. the first, name given to a species should invariably stand in no case (unless preoccupied) be replaced by another name thereafter.

Thus for instance the generic name HICORIA was given by Rafinesque to the Hickories in 1808. To the same genus Nuttall, in 1818, applied the name *Carya*, which has since been used in the Manuals in this country. In this case the name HICORIA is to be retained, *Carya* being cited merely as a synonym. Similarly *Nasturtium* is replaced by RORIPA; *Anemonella* by SYNDESMON; *Desmodium* by MEIBOMIA; *Liatris* by LACINARIA, etc.

The specific names of the plants also have been treated in the same manner, the names retained or restored that properly belong to them. Thus the Kentucky Coffee-tree is called GYMNOCLADUS DIOICUS (L.) Koch. instead of *Gymnocladus canadensis* Lam.* as heretofore printed in the Manuals. The "L." in parenthesis indicates that this specific name was first applied by Linnaeus—though he placed it in the wrong genus (namely *Guilandina*†); the author whose name (or abbreviation) follows the parenthesis is responsible for its present generic status. Similarly the name of the Sugar Maple, ACER SACCHARUM (Marsh.) Britt. replaces *Acer saccharinum* Wang.; for our Silver Maple the name ACER SACCHARINUM L. replaces *Acer dasycarpum* Ehrh.; SASSAFRAS SASSAFRAS (L.) Karst. replaces *Sassafras officinale* Nees; CASTANEA DENTATA (Marsh) Sudw. replaces *Castanea sativa* var. *americana* Wats. & Coult., etc.

The corrected names for all the plants growing in the Northeastern United States can be seen in the "List of the Pteridophyta and Spermatophyta" edited by a committee of the Botanical Club (Amer. Asso. Adv. Sci.) and published by the TORREY BOTANICAL CLUB, Columbia College, New York City. This List has been followed in the printing of the names in this work.

*1783. †1753.

Preface — To Teachers.

It has not seemed best, however, to follow the plan there employed of capitalizing certain specific names. The simpler, and I think the proper method — that now followed by very many if not most leading botanists — of entire decapitalization, has been followed.

To avoid the possibility of any confusion when reference to or comparison with the names as given in GRAY'S REVISED MANUAL (6th edition) is made, the necessary synonyms are given in parentheses.

The "Catalogue of Ohio Plants" by Kellerman and Werner, published in the Geology of Ohio, Vol. VII, part II, pp. 56–406, has been followed in selecting the species that are given in the following pages.

This FLORA is not designed to accompany any particular text-book on botany. It can be used equally well in connection with any one of the numerous botanical text-books found in the various schools throughout the State.

It is suggested, however, that instead of the ordinary plan of text-book use, either the *laboratory method* be employed (in which case Spalding's "Introduction to Botany;" or Arthur, Barnes and Coulter's "Plant Dissection" could be advantageously used); or the instruction be given orally by the Teacher. The main points in the structure, organography and physiology of plants could be given if such a plan were followed. Blackboard outlines, figures and illustrative specimens should be efficient aids. The work could be properly adjusted to the ability and needs of each member of the class. The study could be a real study of the plants, their organs, functions, etc., to the exclusion of redundant terminology. The only book required on the part of the pupils would be this FLORA, by which they could identify the native plants. The Teacher could use for the selection and arrangement of the topics, any one of the more comprehensive text-books, as Bessey's "Essentials of Botany," Bastin's "Elements of Botany," Vine's "Student's Text-book," Behren's "Text-book of General Botany," Kellerman's "Elements of Botany," Gray's botanical works, etc.

Any suggestions from the teachers of the State in reference to an improved edition will be most thankfully received.

W. A. K.

Ohio State University, Columbus.
January, 1895.

PREFACE.=TO TEACHERS.

THIS Florula has been compiled in response to a demand for a book simpler and briefer than the extant Manuals that include the entire flora of a very large portion of the United States.

To place before the pupils descriptions of the plants that do not occur in their region and which therefore they will not see, much less handle, not merely entails a useless expense in the purchase of a large book when a smaller one would answer the purpose, but it also requires him, as it were, to thresh over a large amount of chaff for the few grains obtained.

The object of placing a book with keys, descriptions, etc., in the hands of the pupil should be, to aid him by inciting to farther observation and study of the plants in his neighborhood. Its purpose is to introduce him to the numerous and perhaps hitherto unnoticed plants, to present their names, which will lead in the great majority of cases to a real and protracted acquaintance. This will in time awaken the interesting and important questions as to minute structure, functions, affinities, origin, etc.

The abridgement here presented takes cognizance of the plan, that is quite prevalent or indeed universal in the schools of the State, which provides that the work in elementary botany begin after the Holidays and close the latter part of May or early in June. The plants whose flowers usually appear in this period of time are included. Difficult groups however, as Grasses, Sedges, etc., which beginners never attempt, are excluded.

All the woody plants (trees and shrubs) are included, regardless of their time of flowering, and an analytical key is added for their especial identification based on the characters exhibited by their leaves and fruits.

The generic diagnoses are not exhaustive but enumerate a few and only those characters which most plainly separate the genera as occurring in our flora. The specific characters are likewise referred

Spring Flora of Ohio.

	PAGE		PAGE		PAGE
Cypripedium	26	Hippocastanaceæ	78	Magnoliaceæ	41
Delphinium	44	Houstonia	111	Maianthemum	22
Dentaria	55	Hydrangea	59	Malva	81
Descurainia	56	Hydrastis	43	Malvaceæ	81
Dicentra	51	Hydrophyllaceæ	100	Martynia	109
Diervilla	116	Hydrophyllum	100	Medeola	23
Dioscorea	25	Hypericaceæ	81	Medicago	69
Dioscoreaceæ	25	Hypericum	81	Melilotus	70
Diospyros	95	Hypoxis	25	Menyanthes	97
Dirca	85	Ilex	76	Mertensia	101
Disporum	22	Ilicioides	76	Mitchella	112
Draba	55	Iodanthus	53	Mitella	58
Ebenaceæ	95	Ipomœa	99	Mollugo	37
Elæagnaceæ	85	Iridaceæ	25	Moraceæ	36
Epigæa	92	Iris	25	Morus	36
Ericaceæ	90	Isopyrum	43	Muscari	21
Erigenia	88	Jeffersonia	48	Myosotis	101
Erigeron	119	Juglandaceæ	27	Myrica	29
Erodium	72	Juglans	28	Myricaceæ	29
Erythronium	21	Juniperus	18	Nasturtium	53
Euonymus	76	Kalmia	91	Naumbergia	95
Euphorbia	74	Kraunhia	71	Negundo	78
Euphorbiaceæ	74	Krigia	121	Nemopanthes	76
Fagaceæ	32	Labiatæ	103	Nepeta	103
Fagus	33	Lamium	104	Nigella	48
Ficaria	48	Larix	17	Nuphar	41
Flœrkea	75	Lauraceæ	49	Nymphæa	41
Fragaria	65	Ledum	91	Nymphæaceæ	41
Fraxinus	96	Legouzia	118	Nyssa	90
Galium	112	Leguminosæ	68	Oakesia	20
Gaultheria	92	Lentibulariaceæ	108	Obolaria	97
Gaylussacia	93	Lepargyræa	85	Oleaceæ	96
Gemmingia	25	Lepidium	52	Opulaster	61
Gentianaceæ	97	Ligustrum	97	Orchidaceæ	26
Geraniaceæ	72	Liliaceæ	19	Orchis	26
Geranium	72	Limnanthaceæ	75	Orobanchaceæ	108
Geum	66	Lindera	50	Ornithogalum	21
Glechoma	103	Liquidamber	60	Osmorrhiza	88
Gleditschia	69	Liriodendron	42	Ostrya	31
Gymnocladus	69	Lithospermum	102	Oxalidaceæ	72
Habenaria	27	Lonicera	115	Oxalis	73
Hamamelidaceæ	59	Loranthaceæ	36	Oxydendron	92
Hamamelis	60	Lupinus	69	Panax	86
Helianthemum	82	Lycium	105	Papaveraceæ	50
Hepatica	45	Lysimachia	95	Parthenocissus	80
Heracleum	87	Lythraceæ	85	Passiflora	84
Heuchera	58	Lythrum	85	Passifloraceæ	84
Hibiscus	81	Maclura	36	Pedaliaceæ	109
Hicoria	28	Magnolia	41	Pedicularis	107

Index to Orders and Genera.

Name	PAGE	Name	PAGE	Name	PAGE
Pentstemon	105	Rutaceae	73	Tecoma	109
Phacelia	100	Salicaceae	29	Thalesia	108
Phlox	98	Salix	30	Thalictrum	18
Phoradendron	37	Salvia	103	Thaspium	87
Physiocarpus	61	Sambucus	111	Thelypodium	53
Pimpinella	88	Sanguinaria	50	Thuya	18
Pinus	17	Sanicula	87	Thymelaeaceae	85
Platanaceae	60	Santalaceae	37	Tiarella	58
Platanus	60	Sassafras	49	Tilia	81
Plantaginaceae	110	Saxifraga	58	Tiliaceae	80
Plantago	110	Saxifragaceae	57	Tissa	10
Podophyllum	18	Schollera	91	Toxylon	36
Pogonia	27	Scrophulariaceae	105	Tradescantia	19
Polemoniaceae	98	Scutellaria	103	Trautvetteria	46
Polemonium	98	Senecio	120	Trientalis	95
Polygala	74	Shepherdia	85	Trifolium	70
Polygalaceae	74	Silene	38	Trillium	23
Polygonaceae	37	Simarubaceae	73	Triosteum	115
Polygonatum	22	Sisymbrium	52	Trollius	43
Populus	29	Sisyrinchium	25	Tsuga	18
Portulaca	38	Smilaceae	21	Tussilago	120
Portulacaceae	38	Smilacina	22	Ulmaceae	35
Potentilla	65	Smilax	21	Ulmaria	67
Primulaceae	95	Solanaceae	101	Ulmus	35
Prunella	104	Solanum	105	Umbelliferae	86
Prunus	68	Solea	84	Unifolium	22
Ptelea	73	Sorbus	62	Utricularia	108
Pyrus	62	Spathyema	18	Uvularia	20
Quercus	33	Specularia	118	Vaccinium	93
Ranunculaceae	42	Spergula	40	Vagnera	22
Ranunculus	46	Spiraea	61	Valeriana	117
Rhamnaceae	79	Staphylea	77	Valerianaceae	117
Rhamnus	79	Staphyleaceae	77	Valerianella	117
Rhododendron	91	Stellaria	39	Veronica	106
Rhus	75	Stenophragma	56	Viburnum	114
Ribes	59	Streptopus	22	Vicia	71
Robinia	71	Stylophorum	50	Vinca	97
Roripa	53	Sullivantia	58	Viola	82
Rosa	67	Symphoricarpus	115	Violaceae	82
Rosaceae	60	Symplocarpus	18	Vitaceae	79
Rubiaceae	111	Synandra	104	Vitis	80
Rubus	64	Syndesmon	45	Waldsteinia	66
Rudbeckia	119	Syringa	96	Wistaria	71
Ruellia	110	Taxus	18	Xanthoxylum	73
Rumex	37	Taraxacum	121	Zizia	88

14   *Spring Flora of Ohio.*

                                                                                                   PAGE

- 70. Leaves punctate with translucent and dark dots........*Hypericaceæ*. 81
- 70. Leaves not puctate with dots 71.
    - 71. Sepals 5, very unequal or only 3..................*Cistaceæ*. 82
    - 71. Sepals and petals 4, stamens 6......................*Cruciferæ*. 51
    - 71. Sepals and petals 5; Stamens 5 or 10 72.
  - 72. Ovary and stamens raised on a stalk............*Passifloraceæ*. 84
  - 72. Ovary not raised on a stalk..........................*Saxifragaceæ*. 57
    - 73. Flowers irregular 74.
    - 73. Flowers regular or nearly so 78.
- 74. Anthers opening at top, 6-8, 1-celled; ovary 2-celled..........*Polygalaceæ*. 37
- 74. Anthers opening lengthwise 75.
    - 75. Stamens 12; petals 6, on the throat of the calyx ....*Lythraceæ*. 85
    - 75. Stamens 5-8 or 10; petals nearly or quite hypogynous 76.
- 76. Ovary 3-celled; leaves digitate........................*Hippocastanaceæ*. 78
- 76. Ovary 3-celled; leaves trifoliate.......................*Staphyleaceæ*. 77
- 76. Ovary 5-celled 77,.
    - 77. Flowers 5-merous; leaves more or less lobed or divided..*Geraniaceæ*. 72
    - 77. Flowers 3-merous; leaves pinnate, alternate..........*Limnanthaceæ*. 75
    - 77. Flowers 5-merous; leaves trifoliate; styles 5.............*Oxalidaceæ*. 72
      - 78. Stamens neither just as many, nor twice as many as the petals 79.
      - 78. Stamens just as many or twice as many as the petals 81.
- 79. Stamens united by their filaments in 3 sets; petals 5........*Hypericaceæ*. 81
- 79. Stamens distinct, 4 long and 2 short rarely 2 or 4; petals 4...*Cruciferæ*. 51
- 79. Stamens distinct and fewer than the 4 petals; trees or shrubs....*Oleaceæ*. 96
- 79. Stamens distinct and more numerous than the petals 80.
    - 80. Leaves digitate; flowers irregular..................*Hippocastanaceæ*. 78
    - 80. Leaves simple or pinnate; fruit winged.................*Aceraceæ*. 77
    - 80. Leaves trifoliate; fruit a bladdery inflated pod........*Staphyleaceæ*. 77
      - 81. Ovules only 1 or 2 in each cell 82.
      - 81. Ovules several or many in each cell 86.
- 82. Herbs; flowers monœcious in a cup-like involucre..........*Euphorbiaceæ*. 74
- 82. Herbs; flowers perfect and symmetrical 83.
- 82. Shrubs or trees 84.
    - 83. Flowers 5-merous; leaves more or less lobed or divided..*Geraniaceæ*. 72
    - 83. Flowers 3-merous; leaves pinnate and alternate......*Limnanthaceæ*. 75
    - 83. Flowers 5-merous; leaves trifoliate; styles 5..............*Oxalidaceæ*. 72
      - 84. Leaves palmately veined; fruit 2-winged............*Aceraceæ*. 77
      - 84. Leaves pinnately veined; fruit not winged 85.
- 85. Pods colored, seeds in a pulpy aril......................*Celastraceæ*. 76
- 85. Fruit a berry-like drupe; calyx minute.................*Aquifoliaceæ*. 76
- 86. Stipules between the opposite and the compound leaves, but caducous...*Staphyleaceæ*. 77
- 86. Stipules none when the leaves are opposite 87.
    - 87. Stamens united by their filaments at base, leaflets 3....*Oxalidaceæ*. 72
    - 87. Stamens distinct, free from the calyx 88.
    - 87. Stamens distinct, inserted on the calyx 89.
      - 88. Style undivided..*Ericaceæ*. 90
      - 88. Styles 2-5, separate..................................*Caryophyllaceæ*. 38
- 89. Styles 2 or 3 or splitting into 2 in fruit..............*Saxifragaceæ*. 67

Key to the Orders. 15

PAGE
- 89. Style 1, pod in the calyx and 1-celled at maturity........*Lythraceæ*.* 85
  - 90. Ovules more than one in each cell 91 .
  - 90. Ovules only one in each cell 93 .
    - 91. Ovary 1-celled ; many-ovuled from the base............*Portulacaceæ*. 38
    - 91. Ovary 1-celled ; placentas 2 or 3, parietal...... ....*Saxifragaceæ*. 57
    - 91. Ovary 2-several-celled 92 .
- 92. Stamens on a flat disk which covers the ovary ; shrubs......*Celastraceæ*. 76
- 92. Stamens inserted on the calyx ; styles 2-3, distinct.........*Saxifragaceæ*. 57
  - 93. Stamens 10 or 5 ; fruit a pome ; styles 1-many..............*Rosaceæ*. 60
  - 93. Stamens 4 ; styles and stigmas 1 ; chiefly shrubs........ ...*Cornaceæ*. 89
  - 93. Stamens 5 ; fls. in umbels ; styles 2 ; fruit dry..........*Umbelliferæ*. 86
  - 93. Stamens, etc., as the last ; styles usually more than 2, fruit
    a drupe.......... ....*Araliaceæ*. 85

GAMOPETALOUS GROUP.

- 94. Stamens more numerous than the lobes of the corolla 95 .
- 94. Stamens fertile as many as and opposite the corolla-lobes..*Primulaceæ*. 95
- 94. Stamens fewer than, or as many as, and alternate with the corolla-lobes 99 .
  - 95. Ovary 1-celled, with one parietal placenta. ....*Leguminosæ*. 68
  - 95. Ovary 2-celled, with a single ovule in each cell.........*Polygalaceæ*. 78
  - 95. Ovary 3-many-celled 96 .
- 96. Stamens nearly or quite free from the corolla ; style single....*Ericaceæ*. 90
- 96. Stamens free from the corolla ; styles 5 97 .
- 96. Stamens inserted on the base or tube of the corolla 98 .
  - 97. Leaves more or less lobed or pinnately divided ....*Geraniaceæ*. 72
  - 97. Leaves trifoliate................................ ....*Oxalidaceæ*. 72
- 98. Filaments united into a tube ; anthers 1-celled..................*Malvaceæ*. 81
- 98. Filaments wholly distinct ; calyx free, persistent..............*Ebenaceæ*. 95
  - 99. Ovary adherent to the calyx-tube 100 .
  - 99. Ovary free from the calyx ,103 .
- 100. Stamens united by their anthers ; flowers in heads ; corollas all
    tubular or the marginal ones ligulate................ ..*Compositæ*. 118
- 100. As above, but all the corollas ligulate....................,......*Cichoriaceæ*. 121
- 100. Stamens separate, free or nearly so from the corolla ; juice
    milky ....*Campanulaceæ*. 118
  - 101. Stamens 1-3, fewer than the corolla-lobes........*Valerianaceæ*. 117
  - 101. Stamens 4-5 ; leaves opposite or whorled 102 .
- 102. Leaves opposite or whorled, with or without stipules ....*Rubiaceæ*. 111
- 102. Leaves opposite, without stipules......................*Caprifoliaceæ*. 113
  - 103. Corolla irregular ; stamens with anthers 4 of 2 lengths, or 2 104 .
  - 103. Corolla regular 105 .
- 104. Ovules solitary in the cells ; ovary deeply 4-lobed ....*Labiatæ*. 103
- 104. Ovules 2 or more in each cell 105 .
  - 105. Ovary and pod 1-celled 106 .
  - 105. Ovary and pod 2-celled ; placentas 2, in the axis 107 .
  - 105. Ovary and pod 2-celled ; the two placentas parietal....*Bignoniaceæ*. 108
  - 105. Ovary and fruit more or less 4-5-celled..................*Pedaliaceæ*. 109
- 106. Placenta free, central ; stamens 2 ; plants aquatic........*Lentibulariaceæ*. 108
- 106. Placentas 2 or more, parietal ; many seeded ; stamens 4....*Orobanchaceæ*. 108

12   Spring Flora of Ohio.

35. Ovary 1-celled; style and stigma single and entire; anthers open-   PAGE
  ing by uplifted valves..Lauraceæ. 49
35. Ovary, etc., as in the last; anthers opening longitudinally...Thymelæaceæ. 85
  36. The fruit a winged samara or a drupe......................Ulmaceæ. 35
  36. The calyx becoming fleshy or juicy in fruit.................Moraceæ. 36
  36. The fruit a woody, 2-beaked capsule...............Hamamelidaceæ. 59
37. Ovary 6-celled; stamens 6-12.............................Aristolochiaceæ. 37
37. Ovary 2-celled, 2-beaked; calyx absent; tree....   ......Hamamelidaceæ. 59
37. Ovaries 2 or more, separate, simple.....................Ranunculaceæ. 74 42
  38. Ovary 3-celled and 3-valved................................Aizoidaceæ. 37
  38. Ovary 2-celled, 2-beaked; calyx absent; tree........Hamamelidaceæ. 59
  38. Ovary 2 or 1-celled; placentas central; calyx present 39).
  38. Ovary 1-celled, with 1 parietal placentas...............Ranunculaceæ. 74 42
39. Stamens inserted on the calyx-tube..............................Lythraceæ. 85
39. Stamens on the receptacle or base of the calyx............Caryophyllaceæ. 38

POLYPETALOUS GROUP.

  10. Stamens more than 10; more than twice the number of petals 12).
  10. Stamens not more than twice the number of petals 41.
11. Stamens of the same number as, and opposite the petals 53.
11. Stamens alternate with the petals or of a different number 56.
    12. Calyx entirely free from the pistil or pistils 13.
    12. Calyx more or less coherent with the compound ovary 52.
    13. Pistils numerous but cohering over each other in a solid
      mass or on an elongated receptacle.................Magnoliaceæ. 41
    13. Pistils separate but concealed in a hollow receptacle, leaves
      alternate, with stipules.....................................Rosaceæ. 60
    13. Pistils like the last, leaves opposite, exstipulate.....Calycanthaceæ. 49
    13. Pistils separate, not enclosed in the receptacle 14.
    14. Pistils with their ovaries cohering in a ring around the
      axis...Malvaceæ. 81
    13. Pistils but one as to the ovary; styles or stigmas may be several 47.
  44. Stamens distinct, inserted on the calyx ....Rosaceæ. 60
  44. Stamens united by their filaments, and with the base of the
    petals...Malvaceæ. 81
  44. Stamens inserted on the receptacle 45.
    45. Filaments much shorter than the anthers; trees..........Anonaceæ. 42
    45. Filaments longer than the anthers 46.
  46. Leaves peltate; petals persistent.......................Nymphæaceæ. 41
  46. Leaves not peltate; petals deciduous...................Ranunculaceæ. 74 42
    47. Leaves punctate under a lens, with transparent dots ..Hypericaceæ. 81
    47. Leaves not punctate with transparent dots 48.
48. Ovary simple, 1-celled, 2-ovuled............................Rosaceæ. 60
48. Ovary simple, 1-celled, many-ovuled, placenta parietal 49.
48. Ovary compound, 1-celled, placenta central..................Portulacaceæ. 38
48. Ovary compound, 1-celled, placentas 2 or more, parietal 50.
48. Ovary compound, several-celled 51.
  49. Leaves 2-3-ternately decompound or dissected.........Ranunculaceæ. 74 42
  49. Leaves peltate and lobed..............................Berberidaceæ. 48

Key to the Orders. 13

	PAGE
50. Calyx caducous, juice milky or colored............ ......*Papaveraceæ*.	50
50. Calyx persistent; of 3 or five sepals, juice not as above.........*Cistaceæ*.	82
51. Calyx persistent; stam. monadelphous, anthers 1-celled....*Malvaceæ*.	81
51. Calyx persistent; stamens distinct; aquatic plants.....*Nymphæaceæ*.	41
51. Calyx deciduous; stamens distinct; trees..................*Tiliaceæ*.	80
52. Ovary 10 to 30-celled; ovules many; aquatic plants.... ......*Nymphæaceæ*.	41
52. Ovary 2 to 5-celled; leaves alternate, with stipules................*Rosaceæ*.	60
52. Ovary 1-celled, with the ovules arising from the base.........*Portulacaceæ*.	38
53. Ovary 1-celled, anthers opening by uplifted valves .....*Berberidaceæ*	18
53. Ovary 1-celled; anthers not opening as above 54 .	
53. Ovary 2 to 4-celled 55 .	
54. Style and stigma one.................... ....*Primulaceæ*.	95
54. Style 1, stigmas 3, sepals 2................ ....*Portulacaceæ*.	38
55. Calyx-lobes minute or obsolete, petals valvate.....................*Vitaceæ*.	79
55. Calyx 1 to 5-cleft, valvate in the bud, petals involute....... ......*Rhamnaceæ*.	79
56. Calyx free from the ovary, the latter wholly superior 57 .	
56. Calyx-tube adherent to the ovary or its lower half 90	
57. Ovaries 2 or more, separate 58 .	
57. Ovaries 2-5, separate above, united at base 62 .	
57. Ovaries or the lobes 3-5, with a common style.................*Geraniaceæ*.	72
57. Ovary only one 65 .	
58. Stamens on the receptacle, free from the calyx 59	
58. Stamens inserted on the calyx 61 .	
59. Leaves punctate with pellucid dots......................*Rutaceæ*.	73
59. Leaves not punctate with pellucid dots 60 .	
60. Tree with pinnate leaves................*Simarubaceæ*.	73
60. Herbs ....*Ranunculaceæ*.	12
61. Leaves with stipules ....*Rosaceæ*.	60
61. Leaves without stipules.................................*Saxifragaceæ*.	57
62. Leaves punctate with pellucid dots..........................*Rutaceæ*.	73
62. Leaves not pellucid-punctate 63 .	
63. Shrubs or trees with opposite leaves 64 .	
63. Terrestrial herbs; carpels fewer than the petals.............*Saxifragaceæ*.	57
64. Leaves palmately compound, flowers irregular......*Hippocastanaceæ*.	78
64. Leaves simple or pinnate; fruit winged....................*Aceraceæ*.	77
64. Leaves trifoliate; fruit an inflated, 3-celled pod........*Staphyleaceæ*.	77
65. Ovary simple; placenta one, parietal ....*Leguminosæ*.	68
65. Ovary compound, as shown by cells, placentas, styles or stigmas 66 .	
66. Ovary 1-celled 67 .	
66. Ovary 2-several-celled 73 .	
67. Corolla irregular, petals 4, stamens 6........................*Papaveraceæ*.	50
67. Corolla irregular petals and stamens 5 ....*Violaceæ*.	82
67. Corolla regular or nearly so 68 .	
68. Ovules solitary; stigmas 3, shrubs or trees............*Anacardiaceæ*.	75
68. Ovules solitary or few; herbs ....*Cruciferæ*.	51
68. Ovules more than one, in the center or bottom of the cell 69 .	
68. Ovules several, on two or more parietal placentas 70 .	
69. Petals not inserted on the calyx ....*Caryophyllaceæ*.	38
69. Petals on the throat of the calyx................................*Lythraceæ*.	85

KEY TO THE ORDERS.

Cone-bearing plants, the so-called "Evergreens" 4.
Plants not cone-bearing, leaves not needle-shaped nor awl-shaped 1.
1. Stems endogenous, leaves mostly parallel-veined ; fls. usually 3-parted 5.
1. Stems exogenous, lvs. mostly netted-veined ; fls. usually 5-parted 2.
   2. Calyx and corolla both present 3.
   2. Corolla and sometimes also the calyx wanting 11.
3. Corolla of separate petals 10.
3. Corolla of more or less united petals 94.

GYMNOSPERMOUS PLANTS.

PAGE

4. Trees or shrubs ; leaves scale, awl or needle-shaped..............*Coniferæ*. 17

ANGIOSPERMOUS PLANTS.

MONOCOTYLS.

5. Fls. on a spadix ; lvs. petioled, mostly netted-veined..............*Araceæ*. 18
5. Flowers not on a spadix 6
   6. Perianth adherent to the inferior ovary 7.
   6. Perianth free from the superior ovary 9.
7. Flowers diœcious or polygamous ; stem twining..............*Dioscoreaceæ*. 25
7. Flowers perfect ; stems not twining 8.
   8. Stamens united with the pistils ; fls. irregular............*Orchidaceæ*. 26
   8. Stamens three ; leaves 2-ranked............................*Iridaceæ*. 25
   8. Stamens six ; flowers on a scape from a bulb..........*Amaryllidaceæ*. 24
9. Anthers 1-celled ; fls. diœcious ; plants tendril-bearing..........*Smilaceæ*. 24
9. Anthers 2-celled ; plants not tendril-bearing 10.
   10. Perianth of similar divisions or lobes, mostly colored ; or of 3
      foliaceous sepals and 3 colored, persistent petals.........*Liliaceæ*. 19
   10. Perianth of 3 persistent sepals and 3 ephemeral petals.*Commelinaceæ*. 19

DICOTYLS.

APETALOUS GROUP.

   11. Flowers not in catkins 20.
   11. Flowers, one or both sorts, in catkins 12.
12. Only one sort of flowers in catkins or catkin-like heads 13.
12. Both sterile and fertile flowers in catkins or catkin-like heads 15
   13. Fertile flowers in a short catkin or head.................*Moraceæ*. 36
   13. Fertile flowers, single or clustered, the sterile in catkins 14.

Key to the Orders. 11

14. Leaves pinnate; fruit walnut or hickory-nut naked........*Juglandaceæ.* 27
14. Leaves simple; fertile flowers and fruit in an involucre or cup..*Fagaceæ.* 32
   15. Ovary 1-celled, many-seeded; seeds with a down...........*Salicaceæ.* 29
   15. Ovary 1-2-celled, each cell 1-ovuled; fruit 1-seeded (16).
      16. Parasitic on trees; fruit a berry................*Loranthaceæ.* 36
      16. Trees or shrubs, not parasitic 17
17. Calyx of the fertile flowers succulent in fruit................*Moraceæ.* 36
17. Calyx none or rudimentary or scale-like 18
   18. Style and stigma, one, simple; the flowers in heads.....*Platanaceæ.* 60
   18. Styles or long stigmas two 19.
19. Fertile flowers 2 or 3 under each scale of the catkin or each in a
   sac or with bractlets................................*Betulaceæ.* 31
19. Fertile flowers single under each scale; nutlets naked, waxy-coated
   or drupe-like...*Myricaceæ.* 29
   20. Ovary or its cells containing many ovules 21.
   20. Ovary or its cells containing 1-2 rarely 3-4 ovules 22.
      21. Ovary one and inferior; or 2 or more separate ovaries 37.
      21. Ovary one, superior; or the calyx entirely wanting 38.
      22. Pistils more than one, distinct or nearly so 23.
      22. Pistil one, either simple or compound 25.
23. Stamens on the calyx; leaves with stipules......................*Rosaceæ.* 60
23. Stamens inserted on the receptacle 24.
   24. Leaves punctate with translucent dots.....................*Rutaceæ.* 97
   24. Leaves not dotted; calyx often petaloid...............*Ranunculaceæ.* 42
25. Ovary half inferior, 2-celled; styles 2, pod 2-beaked, woody.*Hamamelidaceæ.* 59
25. Ovary wholly inferior in perfect or pistillate flowers 26.
25. Ovary really free but invested by the calyx tube.............*Rosaceæ.* 60
25. Ovary plainly free from the calyx, or calyx wanting 29.
   26. Aquatic herbs; ovary 4-celled......................*Callitrichaceæ.* 74
   26. Woody plants or herbs; ovary 1-2-celled 27.
27. Stigmas running down one side of style.......................*Cornaceæ.* 89
27. Stigmas terminal, with or without a style 28.
   28. Parasitic on branches of trees; anthers sessile.........*Loranthaceæ.* 36
   28. Not parasitic above ground; anthers on filaments.......*Santalaceæ.* 37
29. Stipules sheathing the stem at the nodes 30.
29. Stipules none, or not sheathing the stems 31.
   30. Tree; calyx none; flowers monœcious in heads..........*Platanaceæ.* 60
   30. Herbs; calyx present, usually petaloid ....*Polygonaceæ.* 37
      31. Aquatic herbs; leaves opposite, entire............*Callitrichaceæ.* 74
      31. Herbs, not aquatic 32.
      31. Shrubs or trees 33.
   32. Ovary 3-celled; juice usually milky, fls. in an involucre.*Euphorbiaceæ.* 74
   32. Ovary 1-celled; juice not milky, stipules scarious....*Caryophyllaceæ.* 34
      33. Ovules a pair or several in each cell of the ovary 34.
      33. Ovules single in each cell of the ovary 35.
34. Fruit a 2-celled, double samara................................*Aceraceæ.* 77
34. Fruit a 1-celled, 1-seeded samara or drupe...... ....*Oleaceæ.* 96
34. Fruit a 2-beaked, woody capsule......................*Hamamelidaceæ.* 59
35. Ovary 3-celled; branches often thorny.......................*Rhamnaceæ.* 79
35. Ovary 1 to 2-celled; styles or stigmas 2-cleft (36).

Spring Flora of Ohio.

PAGE
- 107. Seeds few, borne on hooks or projections of the placentas..*Acanthaceæ*. 103
- 107. Seeds mostly many, not on hooks............. .....*Scrophulariaceæ*. 105
- 108. Stamens as many as the lobes of the corolla 109.
- 108. Stamens fewer than the lobes of the corolla 117.
  - 109. Ovaries 2, separate. filaments distinct............. *Asclepiadaceæ* 98 *Apocynaceæ*. 97
  - 109. Ovary 1, deeply 4-lobed around the style, lvs. alternate..*Boraginaceæ*. 100
  - 109. Ovary 1, not divided nor lobed 110.
- 110. Ovary 1-celled, ovules parietal or on 2 parietal placentas 111.
- 110. Ovary 2-10-celled 112.
  - 111. Leaves entire, or the leaflets entire..................... *Gentianaceæ*. 97
  - 111. Leaves toothed, lobed or pinnately compound...... *Hydrophyllaceæ*. 100
- 112. Stamens free from the corolla or nearly so, style 1............. *Ericaceæ*. 90
- 112. Stamens almost free from the corolla ; style none ....*Aquifoliaceæ*. 76
- 112. Stamens inserted on the tube of the corolla 113.
  - 113. Stamens 4, pod 2-celled, circumscissile............... *Plantaginaceæ*. 110
  - 113. Stamens 5, rarely more 114.
- 114. Fruit of 2 or 4 seed-like nutlets............................. *Boraginaceæ*. 100
- 114. Fruit a few-seeded pod 115.
- 114. Fruit a very many-seeded pod or berry 116.
  - 115. Style 3-cleft or 3-lobed ; seeds small.................. *Portulaceæ*. 98
  - 115. Style single or 2-cleft ; seeds large.............. ......*Convolvulaceæ*. 99
- 116. Styles 2.. *Hydrophyllaceæ*. 100
- 116. Style single.............. ....*Solanaceæ*. 104
  - 117. Stamens 4, two of them shorter than the others........*Acanthaceæ*. 110
  - 117. Stamens 2, rarely 3 118.
- 118. Herbs ; corolla scarious, withering on the pod.......... *Plantaginaceæ*. 110
- 118. Herbs ; corolla rotate or sub-funnel-form, not scarious...*Scrophulariaceæ*. 105
- 118. Shrubs or trees.. ...*Oleaceæ*. 96

Spring Flora of Ohio.

I. Order CO-NIF'-ER-Æ. PINE FAMILY. Trees or shrubs, resinous, leaves mostly evergreen; scale-like, subulate, or needle-shaped; flowers never perfect, the ovules naked; fruit a cone, dry or berry-like.

KEY TO THE GENERA.

Leaves scale-like and adnate, or free and awl-shaped a .
Leaves scattered, linear, 2-ranked, flat, *green both sides* b .
Leaves 2-5 in a cluster, surrounded by a sheath.........................*Pinus*. 1
Leaves fascicled, very many in a cluster, *deciduous*....................*Larix*. 2
Leaves scattered, petioled, flat, *whitened beneath*.....................*Tsuga*. 3
a. Leaves 2-ranked; fruit a cone with a few scales.........................*Thuya*. 4
a. Leaves not 2-ranked; fruit berry-like..................................*Juniperus*. 5
b. A straggling bush; fruit berry-like nearly enclosing the seed ...*Taxus*. 6

1. Genus **PI'-NUS.** — Trees or shrubs with evergreen needle-shaped, semi-cylindrical or triangular leaves, 2-5 in a sheathed fascicle; fruit a cone with woody scales; seeds winged.

Pi'-nus stro'-bus L. **White Pine.** — A large tree with soft slender glaucous leaves in fives; scales of the long cone slightly thickened at the end, without prickle or point.

Pi'-nus rig'-i-da Mill. Pitch Pine. Leaves in threes, 3-5 in. long from short sheaths; scales of the cone with a short stout recurved prickle.

Pi'-nus vir-gin-i-a'-na Mill. (*P. inops* Ait.) **Jersey Pine: Scrub Pine.** Leaves in twos, short, $1\frac{1}{2}$-3 in. long; scales of cone tipped with a straight or recurved prickle; a shrub or tall tree.

2. Genus **LAR'-IX.** Leaves needle-shaped, soft, *deciduous*, many in a fascicle; cones $\frac{1}{2}$-$\frac{3}{4}$ in. long, scales few, rounded.

Lar'-ix lar-i-ci'-na (Du Roi) Koch. (*L. americana* Mx.) **Tamarack: American Larch: Black Larch.** — A slender tree, growing chiefly in swamps. Northward.

3. Genus **TSU'-GA.** — Leaves scattered, flat, whitened beneath, somewhat 2-ranked; cones pendulous, scales thin, persistent.

Tsu'-ga can-a-den'-sis (L.) Carr. **Hemlock; Hemlock Spruce; Spruce.** — Leaves petioled, short, linear, obtuse, ½ in. long; cones oval, of few thin scales.

4. Genus **THU'-YA.** — Leaves small, closely imbricated, persistent, awl-shaped on some branchlets, on others scale-like, blunt, short and adnate; branchlets very flat; cone small, scales few.

Thu'-ya oc-ci-den-ta'-lis L. **Arbor Vitæ.** — Leaves appressed, imbricated, in four rows on the 2-edged branchlets; scales of the cones pointless.

5. Genus **JU-NIP'-ER-US.** — Trees or shrubs with awl-shaped or scale-like leaves; cones *fleshy*, of 3–6 coalescent scales, when mature berry-like, bluish-black with white bloom.

Ju-nip'-er-us com-mu'-nis L. **Common Juniper.** Shrub or small tree; leaves in whorls of three, linear, subulate, rigid ⅓–¾ in. long.

Ju-nip'-er-us vir-gin-i-a'-na L. **Red Cedar: Savin.** Upright shrub or tall tree; leaves mostly opposite, some scale-shaped, others awl-shaped.

6. Genus **TAX'-US.** — Leaves flat, rigid, mucronate; fruit a nut-like seed, nearly enclosed by a red fleshy berry-like cup.

Tax'-us mi'-nor (Mx.) Britt. (*T. canadensis* Willd.) **American Yew; Ground Hemlock.** — A low irregularly spreading bush; leaves linear, green both sides.

II. Order **A-RA'-CE-Æ. ARUM FAMILY.** — Plants with acrid juice; leaves simple or compound, veiny; flowers on a spadix subtended by a spathe.

Flowers perfect, covering the globular spadix......... *Spathyema.* 1
Flowers monœcious or diœcious, only on base of spadix......... *Arisæma.* 2

1. Genus **SPATH-Y-E'-MA.** (*Symplocarpus.*) — Plants with skunk-like odor, leaves very large, broad, entire, veiny, preceded in early spring by the nearly sessile spathe.

Spath-y-e'-ma fœ'-ti-da (L.) Raf. (*Symplocarpus fœtidus* Salisb.) *Skunk cabbage.* Spathe purple spotted and striped, incurved at the apex.

Monocotyls or Endogenous Plants.

2. Genus **AR-I-SAE′-MA.** — Spathe mostly arched above; leaves divided; scape simple, from a very pungent tuberous root-stock or corm.

Ar-i-sæ′-ma tri-phyl′-lum (L.) Torr. **Indian Turnip.** — Leaves mostly 2, leaflets 3; spadix club-shaped, obtuse, shorter than the spathe.

Ar-i-sæ′-ma dra-con′-ti-um (L.) Schott. **Green Dragon; Dragon-root; Indian Turnip.** Leaf usually solitary, pedately divided into 7–11 leaflets; spadix tapering to a long and slender point.

III. Order **COM-ME-LI-NA′-CE-Æ. SPIDERWORT FAMILY.** — Stems jointed, leafy; sepals 3, persistent; petals 3, ephemeral; stamens 6; ovary 2–3-celled, superior.

1. Genus **TRA-DES-CAN′-TI-A.** — Sepals 3, persistent, herbaceous, petals 3, ephemeral; stems mucilaginous, leafy, the leaves keeled.

Tra-des-can′-ti-a vir-gin-i-a′-na L. **Common Spiderwort.** — Plant more or less glaucous; leaves linear.

Tra-des-can′-ti-a pi-lo′-sa Lehm. (*T. virginica* var. *flexuosa* Wats.) **Hairy Spiderwort.** — Stout and dark green; leaves large, linear-lanceolate, *pubescent*.

IV. Order **LIL-I-A′-CE-Æ. LILY FAMILY.** — Flowers perfect, regular, stamens 6, opposite the segments; ovary 3-celled, superior.

    Flowers in spikes, racemes, corymbs or panicles a .
    Flowers in umbels e .
    Flowers solitary c .
    Flowers terminal, few or many, not as above d .
    Flowers axillary i .
a. Plants with radical leaves, flowers on scapes g .
a. Plants with leafy stems b .
  b. Flowers perfect, very large, segments recurved or spreading f .
  b. Flowers perfect, small, white, not spike-like k .
  b. Flowers diœcious, small, white, in a spike-like raceme...*Chamælirium.* 1
c. Leaves oblong, alternate, sessile-clasping or perfoliate............*Uvularia.* 2
c. Leaves three in a whorl at the summit of the stem n .
c. Leaves in 2 whorls, 3–9 in each whorl m .
c. Leaves 2, elliptic-lanceolate, shining, mostly mottled f .
c. Leaves *ovate*, closely sessile, *plants downy* l .
  d. Leaves oblong, alternate, sessile-clasping or perfoliate [c, above .
  d. Leaves *ovate*, closely sessile, *plants downy* l .
  d. Leaves in 2 whorls of 3–9 each m .
e. Perianth 6-parted, odor alliaceous.............................*Allium.* 3
  f. Stem scape with a pair of mottled leaves, flower nodding.*Erythronium.* 4

g. Leaves 2, oblong, long sheathing petioles, raceme 1-sided m .
g. Leaves linear, flowers blue or greenish-white h .
    h. Flowers light-blue, 6-parted, long racemose..................*Camassia.* 5
    h. Flowers greenish-white, 6-parted, sub-corymbose........*Ornithogalum.* 6
    h. Flowers deep-blue, small, urn-shaped, in a dense raceme......*Muscari.* 7
i. Much branched, the thread-like branchlets appear as leaves........*Asparagus.* 8
    k. Perianth 6-parted, stamens 6, ovary 3-celled............ .....*Vagnera,* 9
    k. Perianth 4-parted, stamens 4, ovary 2-celled.................*Unifolium.* 10
l. Flowers axillary, *on bent pedicels*, anthers sagittate, acute.........*Disporum.* 11
l. Flowers few in umbels [or solitary , anthers oblong, obtuse........*Streptopus.* 12
    m. Stems leafy, flowers axillary, perianth cylindrical........*Polygonatum.* 13
    m. Leaves 2, oblong, raceme, 1-sided, flowers bell-shaped.. ....*Convallaria.* 14
    m. Leaves in 2 whorls, flowers umbellate or solitary..............*Medeola.* 15
n. Leaves 3 in a whorl at the summit of the stem......................*Trillium.* 16

1. Genus **CHAM-Æ-LIR´-I-UM.**—Smooth, stem rather slender, from a bitter thick root-stock; spiked raceme 4–12 in. long.

Cham-æ-lir´-i-um lu´-te-um (L.) Gr. (*C. carolinianum* Willd.) **Blazing Star; Devil's-bit.**— Leaves lanceolate, the lowest spatulate; stem 1–4 ft.

2. Genus **U-VU-LA´-RI-A.**—Stems from a root-stock; leaves sessile-clasping or perfoliate; perianth narrow campanulate.

U-vu-la´-ri-a per-fo-li-a´-ta L. **Bellwort.**—Glaucous throughout, stem terete; 1–3 leaves below the fork, perfoliate, oblong or ovate-lanceolate, glabrous.

U-vu-la´-ri-a gran-di-flo´-ra Sm. **Bellwort.**—Not glaucous; stem terete, naked or a single leaf below the fork; leaves as in the last, but whitish pubescent beneath.

U-vu-la´-ri-a ses-sil-i-fo´-li-a L. (*Oakesia sessilifolia* Wats.) **Bellwort.**- Stem acutely angled; leaves lance-oblong, acute at each end, pale, glaucous beneath, sessile or partly clasping.

3. Genus **AL´-LI-UM.** Strong scented and pungent, stemless herbs; the leaves and scapes from a bulb; spathe scarious, 1–3 valved.

    A. *Leaves elliptic-lanceolate.*

Al´-li-um tri-coc´-cum Ait. **Wild Leek.** -Scape 4–12 in. high, umbel erect, many-flowered; leaves 5–9 in. long, 1–2 in. wide, appearing in early spring and dying before the flowers are developed.

    A. *Leaves linear, flat, channeled or terete.*

Monocotyls or Endogenous Plants.

Al'-li-um cer'-nu-um Roth. **Onion.** Scape angular, $1/2$–2 ft. high, nodding at the apex, sepals oblong-ovate, acute, rose-color; capsule crested.

Al'-li-um can-a-den'-se L. **Wild Garlic.** — Scape 1 ft. high or more, umbel *densely bulbiferous or few flowered;* sepals narrowly lanceolate.

Al'-li-um vin-e-a'-le L. **Field Garlic.** — Scape slender, 1–3 ft. high; *leaves terete and hollow;* umbel often bulbiferous.

4. Genus **ER-Y-THRO'-NI-UM.** — Nearly stemless; leaves 2, smooth, shining, flat, elliptic-lanceloate, tapering into a petiole, usually mottled; scape 1 flowered, from a deep bulb.

Er-y-thro'-ni-um a-mer-i-ca'-num Ker. **Yellow Adder's-Tongue.** — Leaves mottled with purplish and whitish; perianth light yellow, stigmas united.

Er-y-thro'-ni-um al'-bi-dum Nutt. **White Dog's-tooth Violet.** Leaves less or not at all mottled; perianth pinkish-white; stigmas spreading.

5. Genus **CA-MAS'-SI-A.** Scape and linear keeled leaves from a coated bulb; flowers pale-blue, in a raceme.

Ca-mas'-si-a fra'-se-ri (Gr.) Torr. **Wild Hyacinth: Eastern Camass;** **Quamash.** Scape 1 ft. high or more; bracts longer than the pedicels.

6. Genus **OR-NI-THOG'-A-LUM.** — Scape and linear channeled leaves from a coated bulb; flowers corymbed, perianth of 6 white spreading sepals.

Or-ni-thog'-a-lum um-bel-la'-tum L. **Star of Bethlehem.** Scape 4–9 in. high; flowers on long spreading pedicels. Escaped from gardens.

7. Genus **MUS-CA'-RI.** Leaves and early scape from a coated bulb; flowers small in a dense raceme, perianth urn-shaped or globular.

Mus-ca'-ri bo-try-oi'-des (L.) Mill. **Grape Hyacinth.** — Leaves linear, 3–4 lines broad. Escaped from gardens.

8. Genus **AS-PAR'-A-GUS.** Stems much branched from thick matted root-stocks; flowers small, greenish, axillary, branchlets (called leaves) thread-like, the true leaves being small scales.

As-par'-a-gus of-fic-i-na'-lis L. **Garden Asparagus.** Frequently escaped from gardens.

Spring Flora of Ohio.

9. Genus **VAG'-NE-RA**. Stems simple, from a creeping or thickish root-stock, leaves alternate, mostly sessile; flowers 6-parted, white, in simple or branched racemes.

Vag'-ne-ra ra-ce-mo'-sa (L.) Mor. (*Smilacina racemosa* Desf.) **False Soloman's Seal ; False Spikenard.**—Flowers in a terminal racemose panicle; leaves oblong or oval-lanceolate, acuminate, ciliate.

Vag'-ne-ra stel-la'-ta (L.) Mor. (*Smilacina stellata* Desf.) **False Soloman's Seal.**—Flowers in a simple few-flowered raceme; leaves 7–11, oblong-lanceolate.

Vag'-ne-ra tri-fo'-li-a (L.) Mor. (*Smilacina trifolia* Desf.) **False Soloman's Seal.**—Flowers as in the last; plant dwarf (2–6 in. high); leaves 3 (or 2–4) tapering to a sheathing petiole.

10. Genus **U-NI-FO'-LI-UM**. Like the last genus but the perianth 4-parted, stamens 4, stigmas 2-lobed.

U-ni-fo'-li-um can-a-den'-se (Desf.) Greene. (*Maianthemum canadense* Desf.) **Two-leaved Soloman's Seal.**—Plant 3–5 in. high, leaves 2 or 3, lanceolate to ovate, cordate at base, sessile or shortly petiolate.

11. Genus **DIS'-PO-RUM**. Plants downy, branched above, having closely sessile ovate thin leaves; flowers greenish-yellow, drooping, solitary or in pairs.

Dis'-por-um lan-u-gin-o'-sum (Mx.) Britt. **Downy Disporum.**—Sepals linear-lanceolate, taper-pointed, about $\frac{1}{2}$ in. long.

12. Genus **STREP'-TO-PUS**.—Herbs with forking and divergent branches, ovate and taper-pointed clasping leaves and small flowers, solitary or in pairs on slender *pedicels bent or contorted* near the middle.

Strep'-to pus am-plex-i-fo'-li-us (L.) DC. **Twisted Stalk.**—Leaves very smooth, glaucous underneath; flowers greenish-white; stigma entire.

13. Genus **PO-LYG-O-NA'-TUM**. Herbs with simple erect or curving stems, naked below from creeping thick and knotted root-stocks; leaves alternate; flowers axillary, nodding, greenish, perianth cylindrical, oblong.

Po-lyg-o-na'-tum bi-flo'-rum (Walt.) Ell. **Soloman's Seal.**—Stem 1–3 ft. high, peduncles mostly 2-flowered; perianth 4–6 lines long.

Monocotyls or Endogenous Plants.

Po-lyg-o-na′-tum bi-flo′-rum com-mu-ta′-tum (R. and S.) Mor. **Great Soloman's Seal.** Stem mostly tall (2–6 ft.), peduncles 2 to 8-flowered; perianth 5–9 lines long.

14. Genus **CON-VAL-LA′-RI-A.**—Plant low, stemless, leaves 2, oblong, their long petioles enrolled one within the other; scape angled; raceme 1-sided; flowers white, nodding, sweet-scented.

Con-val-la′-ri-a ma-ja′-lis L. **Lily of the Valley.**—Cultivated, perhaps occasionally escaped.

15. Genus **ME-DE′-O-LA.**—Stem simple, 1–3 ft. high, from a horizontal tuberous white root-stock; leaves 5–9 in a whorl near the middle, also a whorl of usually 3 at the summit.

Me-de′-o-la vir-gin-i-a′-na L. **Indian Cucumber-root.**— Flowers in a sessile umbel, or solitary, recurved.

16. Genus **TRIL′-LI-UM.**—Low stout plants with a whorl of 3 leaves at the summit, netted-veined; flower solitary.

  A. *Ovary 6-angled and more or less winged.*
    b. *Flower sessile.*

Tril′-li-um ses′-si-le L. **Sessile Trillium.** — Leaves sessile, ovate or rhomboidal; sepals spreading, petals sessile.

Tril′-li-um re-cur-va′-tum Beck. **Reflexed Trillium.**—Leaves contracted below into a petiole, ovate-oblong or obovate; *sepals reflexed*, petals pointed, the base narrowed into a claw.

    b. *Flower pedicelled.*

Tril′-li-um e-rec′-tum L. **Purple Trillium.** — Pedicel longer than the flower, usually more or less inclined or declinate; petals ovate to lanceolate, brown-purple, or often white or greenish; stigmas distinct, spreading or recurved.

Tril′-li-um gran-di-flo′-rum (Mx.) Salisb. **Large-flowered Trillium.** — Pedicel longer than the flower, erect or ascending; petals oblanceolate, white, turning rose-color; stigmas erect, somewhat cohering.

Tril′-li-um cer′-nu-um L. **Nodding Trillium.**— Pedicel short, recurved or strongly declinate; petals white or pink, ovate to oblong-lanceolate, wavy, recurved-spreading; stigmas stout, recurved.

  A. *Ovary 3-lobed or angled not winged.*

Spring Flora of Ohio.

Tril'-li-um ni-va'-le. Riddell. **Snowy Trillium.**—Leaves oval or ovate, obtuse; petals oblong, obtuse, white, equalling the peduncle.

Tril'-li-um un-du-la'-tum Willd. (*T. erythrocarpum* Mx.) **Painted Trillium.**—Leaves ovate, taper-pointed; petals ovate or oval-lanceolate, pointed, white, painted at the base with purple stripes, shorter than the peduncle.

V. Order **SMI-LA'-CE-Æ. GREEN BRIER FAMILY.**—Mostly shrubby plants; leaves ribbed and netted-veined, simple; a pair of tendrils on the petiole; flowers small, diœcious, greenish; ovary inferior.

1. Genus **SMI'-LAX.**—Flowers small, greenish, regular, the perianth segments distinct; stigmas thick and spreading, almost sessile.

A. *Stems herbaceous, not prickly, flowers carrion-scented.*

Smi'-lax her-ba'-ce-a L. **Carrion-flower.** Stem somewhat climbing, 3-15 ft. high, *with tendrils;* peduncles elongated, 3-4 in. or more in length, umbel 20 to 40-flowered.

Smi'-lax e-cir-rha'-ta (Englm.) Wats. **Carrion-flower.** Erect 1_2-3 ft. high, *without tendrils* or only the uppermost leaves tendril-bearing; peduncles about equalling the petioles, 1-2$\frac{1}{2}$ in. long.

A. *Stems woody, often prickly.*

b. *Peduncles longer than but seldom twice the length of the petiole.*

Smi'-lax ro-tun-di-fo'-li-a L. **Green Brier.**—Leaves green both sides, ovate or round-ovate, peduncles shorter or scarcely longer than the petioles; branchlets more or less 4-angular.

Smi'-lax glau'-ca Walt. **Green Brier.**—Leaves *glaucous beneath*, ovate, abruptly mucronate, peduncles longer than the short petiole.

Smi'-lax bo'na-nox L. **Green Brier.**—Leaves varying from round-cordate or slightly contracted above the dilated base to fiddle-shaped and halbred-shaped or 3-lobed, green and shining on both sides.

b. *Peduncles 2 to 4 times as long as the petioles.*

Smi'-lax his'-pi-da Muhl. **Green Brier.** Stem below beset with long and weak blackish bristly prickles; leaves membranous and deciduous; peduncles 1$\frac{1}{2}$-2 inches long.

VI. Order **AM-A-RYL-LI-DA'-CE-Æ. AMARYLLIS FAMILY.**—Flowers regular, stamens 6; leaves linear, flat, radical; ovary 3-celled.

Monocotyls or Endogenous Plants.

1. Genus **HY-POX′-IS.** — Perianth persistent, 6-parted; flowers 1–4, yellow inside, umbellate; leaves linear, grassy and hairy.

Hy-pox′-is hir-su′-ta (L.) Willd. (*H. erecta* L.) **Star Grass.** A low grass-like plant with conspicuous flower.

VII. Order **DI-OS-CO-RE-A′-CE-Æ. YAM FAMILY.** Flowers small, stamens 6; pod 3-celled, 3-winged, seeds 1 or 2 in each cell.

1. Genus **DI-OS-CO-RE′-A.** — Flowers diœcious, in axillary panicles or racemes; capsule 3-celled, 3-winged.

Di-os-co-re′-a vil-lo′-sa L. **Wild Yam-root.** Slender, twining over bushes, leaves cordate, 9 to 11-ribbed.

VIII. Order **I-RI-DA′-CE-Æ. IRIS FAMILY.** Flowers perfect; ovary inferior, 3-celled, stamens 3, style single, stigmas 3.

Perianth regular, stigmas not petaloid a.
Outer divisions recurved, inner erect, stigmas petaloid...............*Iris.* 1
a. Stigmas dilated; filaments distinct; rhizome creeping...........*Gemmingia.* 2
a. Stigmas thread-like; filaments united; root fibrous............*Sisyrinchium.* 3

1. Genus **I′-RIS.** Perianth 6-cleft, stamens distinct, the anthers covered by the over-arching petaloid branches of the style.

I′-ris ver-sic′-ol-or L. **Common Blue Flag.** Flowers violet-blue, variegated; stem leafy, 1–3 ft. high, stout, angled on one side; leaves ¾ in. wide.

I′-ris cris-ta′-ta Ait. **Crested Dwarf Iris.** — Stems 3 6 in. high; the outer perianth segments *crested*, but beardless.

2. Genus **GEM-MIN′-GI-A.** Perianth segments widely spreading, orange-yellow, mottled above with purple spots, stamens united only at the base.

Gem-min′-gi-a chi-nen′-sis (L.) Ker. (*Belamcanda chinensis* Adans.) **Blackberry Lily.** — Occasionally escaped from cultivation.

3. Genus **SIS-Y-RIN′-CHI-UM.** Perianth 6-parted; stamens united; leaves grassy or lanceolate; stems 2-edged or winged; flowers small, blue, purplish (or white), from a two-leaved spathe.

Sis-y-rin′-chi-um ber-mu-di-a′-na L. (*S. angustifolium* Mill. and *S. anceps* Cav.) **Blue-eyed Grass.** — In moist meadows, common.

IX. Order **OR-CHI-DA'-CE-Æ. ORCHIS FAMILY.**— Flowers perfect, irregular; ovary inferior, ovules innumerable; pistil and stamens united into a column; pollen usually in masses with a pedicel; the lower modified petal is called the *lip*.

 Plants leafless, brownish, yellowish, or purplish c.
 Plants with one or two leaves, not evergreen b.
 Plants with one leaf, persistent through the winter d.
 Plants with leafy stems a.
a. Flowers solitary or terminal or a few axillary, lipcrested b.
a. Flowers terminal, the lip a very large inflated sac..............*Cypripedium*. 1
 b. Flowers in spikes, showy; *fleshy fibrous roots*.................*Orchis*. 2
 b. Flowers in spikes and otherwise like the preceding except that the two glands or viscid disks on the column formed by the union of the pistil and stamens are not enclosed in a common pouch, but are approximate or widely separated......................*Habenaria*. 3
 b. Flowers solitary, terminal or a few axillary, *root fibrous* or with oblong tubers..*Pogonia*. 4
c. Root-stocks much-branched and toothed coral-like..............*Corallorhiza*. 5
d. The 3 or 4 corms connected horizontally ....*Aplectrum*. 6

1. Genus **CYP-RI-PE'-DI-UM.** Lip a very large inflated sac; leaves large, many-nerved and plaited, sheathing at base.

 A. *Stems leafy.*

  Cyp-ri-pe'-di-um can'-di-dum Willd. **White Lady's Slipper.**— Lip white, striped with purple inside.

  Cyp-ri-pe'-di-um par-vi-flo'-rum Salisb. **Yellow Lady's Slipper.**— Lip bright yellow, flattish from above, 1 in. long or less.

  Cyp-ri-pe'-di-um hir-su'-tum Mill. (*C. pubescens* Willd.) **Larger Yellow Lady's Slipper.** Lip pale yellow, flattened laterally, very convex and gibbous above, 1½–2 in. long.

 A. *Scape naked, 2-leaved at base.*

  Cyp-ri-pe'-di-um a-cau'-le Ait. **Stemless Lady's Slipper.** Scape 8–12 in. high with a green bract at top; lip rose-purple, rarely white, drooping, a fissure in front.

2. Genus **OR'-CHIS.**—Flowers ringent, showy, in a spike, the lip turned downward; the two glands or sticky disks of the stigma in a pouch or hooded fold.

  Or'-chis spec-tab'-il-is L. **Showy Orchis.**—Leaves 2, oblong obovate; the lip undivided.

Monocotyls or Endogenous Plants.

Or'-chis ro-tun-di-fo'-li-a Ph. **Round-leaf Orchis.**—Leaf 1 at the base of stem, orbicular or oblong; the lip 3-lobed.

3. Genus **HAB-E-NA'-RI-A.**—This includes a large number of species of **Rein Orchis**, some of them flowering in June, but too difficult for the beginner to identify, therefore not enumerated.

4. Genus **PO-GO'-NI-A.**—Lip crested or 3-lobed; stems ½-1 ft. high; leaves few or small.

Po-go'-ni-a o-phi-o-glos-soi'-des (L.) Kerr. **Adder's-tongue Pogonia.**—Stem 6-9 in. high, an oval or lance-oblong leaf near the middle and a smaller one or bract near the terminal flower, rarely with 1 or 2 others with a flower in the axil.

Po-go'-ni-a tri-an-thoph'-o-ra (Sw.) B. S. P. (*P. pendula* Lindl.) **Drooping Pogonia.**—Stem 3-8 in. high; leaves 3-4, ¼-½ in. long, alternate, ovate, clasping; flowers drooping, in upper axils.

Po-go'-ni-a ver-ti-cil-la'-ta (Willd.) Nutt. **Whorled Pogonia.**—Stem 6-12 in. high, with scales at the base and a whorl of mostly 5 sessile leaves at the summit.

5. Genus **CO-RAL-LO-RHI'-ZA.**—Flower gibbous or somewhat spurred; plants destitute of green foliage.

Co-ral-lo-rhi'-za co-ral-lo-rhi'-za (L.) Karst. (*C. innata* R. Br.) **Coral-root.** Lip somewhat hastately 3-lobed above the base, flowers 5-12.

Co-ral-lo-rhi'-za o-don-to-rhi'-za (Willd.) Nutt. **Coral-root.** Lip entire or nearly denticulate, with a claw-like base, flowers 6-20.

Co-ral-lo-rhi'-za mul-ti-flo'-ra Nutt. **Coral-root.** Lip deeply 3-lobed, flowers 10-30. This species blooms later.

6. Genus **A-PLEC'-TRUM.** Lip free, 3-ridged in the palate, not spurred or saccate; leaf single, produced in autumn but persisting through the winter.

A-plec'-trum spi-ca'-tum (Walt.) B. S. P. (*A. hyemale* Nutt.) **Putty-root; Adam and Eve.** Flowers rather large, dingy, raceme loose.

X. Order **JUG-LAN-DA'-CE-Æ. WALNUT FAMILY.**—Trees; leaves alternate, pinnate; sterile flowers in catkins, the fertile single or clustered.

Fruit with an indehiscent pericarp; pith in plates........ ......*Juglans.* 1
Fruit with a 4-valved dry exocarp..*Hicoria.* 2

1. Genus **JUG'-LANS.** Sterile flowers in lateral catkins from the wood of the preceding year; stamens 12-40; fertile flowers solitary or several at the end of the branches.

Jug'-lans cin-e'-re-a L. **Butternut: White Walnut.** Leaflets 5-8 pairs, downy especially beneath; petioles and branchlets downy with clammy hairs; fruit oblong.

Jug'-lans ni'-gra L. **Black Walnut.** Leaflets 7-11 pairs, the petioles and underside of leaflets minutely downy; fruit spherical.

2. Genus **HIC-O'-RI-A.** (*Carya.*) Sterile flowers in slender lateral clustered catkins, stamens 3-10; fertile flowers 2-5 in a cluster or short spike at the end of the branch.

A. *Husk of fruit thick and woody, splitting promptly.*
  b. *Bark shaggy or exfoliating in strips or plates.*

Hic-o'-ria o-va'-ta (Mill.) Britt. (*Carya alba* Nutt.) **Shell-bark** or **Shag-bark Hickory.** Leaflets 5-7, the lower pair much smaller; fruit globular or depressed, nut white, flattish-globular, the shell thinnish.

Hic-o'-ria la-cin-i-o'-sa (Mx. f.) Sarg. (*Carya sulcata* Nutt.) **Big Shell-bark: Kingnut.** Leaflets 7-9; fruit oval or ovate, the husks very thick; nut large, $1\frac{1}{4}$-2 in. long, usually angular, dull-white or yellowish, thick-walled.

  b. *Bark close, not exfoliating.*

Hic-o'-ria alba (L.) Britt. (*Carya tomentosa* Nutt.) **Mocker-nut: White-heart Hickory: Black Hickory.** Catkins, shoots and lower surface of leaves *tomentose* when young, resinous scented; leaflets 7-9; fruit globular or avoid, with very thick and hard husk; nut globular, not compressed, 4-ridged toward summit.

Hic-o'-ri-a mi-cro-car'-pa (Nutt.) Britt. (*Carya microcarpa* Nutt.) **White Hickory.** Bark, buds and foliage like the next, fruit small, subglobose with rather thin husk, nut thin-shelled, not angled.

A. *Husk thin, 4-valved to middle or tardily to near base; bark not exfoliating.*

Hic-o'-ria glab'-ra (Mill.) Britt. (*Carya porcina* Nutt.) **Pig-nut: Brown Hickory.** Shoots, catkins and leaves *glabrous* or nearly so, leaflets 5-7, fruit pear-shaped, oblong or oval; nut oblong or oval, $1\frac{1}{2}$-2 in. long, with a thick, bony shell.

Dicotyls or Exogenous Plants.

Hic-o'-ri-a min'-i-ma (Marsh.) Britt. (*Carya amara* Nutt.) **Bitternut: Swamp Hickory.**—Buds yellowish, leaflets 7–11, lanceolate; fruit globular, narrowly 6-ridged; nut globular, short pointed, white, thin-walled.

XI. Order **MYR-I-CA'CE-Æ. SWEET GALE FAMILY.**—Shrubs; flowers in short scaly catkins; leaves resinous, sometimes fragrant; fruit a nut-like drupe.

Leaves pinnatifid with many rounded lobes... ....*Comptonia.* 1
Leaves entire, or somewhat serrate.................. ....*Myrica.* 2

1. Genus **COMP-TO'-NI-A.**—Low shrub; fertile catkins globular, ovary surrounded by 8 linear subulate persistent scales.

Comp-to'-ni-a per-e-gri'-na (L.) Coult. (*Myrica asplenifolia* Endl.) **Sweet Fern.**—Leaves sweet scented, fern-like, linear-lanceolate, stipules semi-cordate.

2. Genus **MY-RI'-CA.**—Shrub, 3–8 ft. high; fertile catkins ovoid; ovary with 2–4 scales at the base.

My-ri'-ca ce-rif'-er-a L. **Bayberry: Wax Myrtle.**—Leaves oblong lanceolate, somewhat preceding the flowers, fragrant; sterile catkins scattered.

XII. Order **SAL-I-CA'-CE-Æ. WILLOW FAMILY.** Trees or shrubs; flowers in catkins; fruit a 1-celled, 2-valved pod; seeds with silky down.

Bracts lacerate, stamens numerous, stigmas elongated...... ..... *Populus.* 1
Bracts entere, stamens few, stigmas short .... *Salix.* 2

1. Genus **POP'-U-LUS.**—Flowers with a cup-shaped disk, obliquely lengthened in front; stamens 8–30 or more; buds scaly; trees with broad leaves.

A. *Styles with 2 or 3 narrow filiform lobes.*
  b. *Petioles laterally flattened; bracts silky.*

Pop'-u-lus al'-ba. White Poplar.—Younger branches and under surface of the oval sinuate-toothed, acute leaves, white tomentose. Cultivated, sometimes escaped.

Pop'-u-lus trem-u-loi'-des Mx. **American Aspen.**—Leaves roundish cordate, teeth small, somewhat regular; scales cut into 3–4 linear divisions, fringed with long hairs.

Pop'-u-lus gran-di-den-ta'-ta Mx.   **Large-toothed Poplar.**—Leaves roundish ovate with large and irregular sinuate teeth; scales cut into 5 or 6 unequal small divisions, slightly fringed.

  b. *Petioles terete, bracts not silky.*

Pop'-u-lus het-er-o-phy'l-la L.   **Downy Poplar: Swamp Poplar.**—Leaves ovate, crenate, white woolly when young, at length nearly smooth.

 A. *Styles with dilated lobes.*

Pop'-u-lus bal-sam-if'-er-a L.   **Balsam Poplar: Tacamahac.**—Buds covered with fragrant resin; leaves ovate-lanceolate, gradually tapering and pointed, crenate; petioles terete; stamens 20-30.

Pop'-u-lus bal-sam-if'-er-a can'-di-cans (Ait.) Gray.   **Balm-of-Gilead.**—Like the last, but the leaves broader and more or less cordate; petiole commonly hairy.   Cultivated.

Pop'-u-lus mo-nil-if'-er-a Ait.   **Cottonwood: Carolina Poplar.**—Leaves broadly deltoid, acuminate, with crenate serratures; petioles flattened, stamens 60 or more.

2. Genus **SA'-LIX.**—Flowers with small glands, stamens mostly 2 (2-10); buds with a single scale; trees or shrubs. The species are very numerous and extremely difficult to determine; only a few of the commoner ones here given.

 A. *Catkins on short, lateral leafy branches.*

  b. *Leaves closely serrate with inflexed teeth.*

Sa'-lix ni'-gra Marsh.   **Black Willow.**—Leaves narrowly lanceolate, very long attenuate from near the roundish or acute base to the usually curved tip, *green both sides.*   Common and variable.

Sa'-lix a-myg-da-loi'-des Anders.   **Amygdaloid Willow.**—Leaves lanceolate or ovate-lanceolate, 2-4 in. long, attenuate cuspidate, *pale or glaucous beneath.*

Sa'-lix al'-ba L.   **White Willow.**—Leaves ashy-gray or silk-white on both sides except when old. The variety *vitellina* (L.) Koch, with yellow twigs, is very common.

Sa'-lix bab-y-lon'-i-ca Tourn.   **Weeping Willow.**—Commonly cultivated; easily recognized by its pendulous branches and linear lanceolate leaves, glaucous beneath.

  b. *Leaves remotely denticulate with projecting teeth.*

Sa'-jix lon-gi-fo'-li-a Muhl. **Long-leaved Willow.** Leaves linear-lanceolate 2-4 in. long, tapering to each end, nearly sessile; catkins linear cyndrical; shrub, along streams.

  b. *Leaves sharply serrate, finely denticulated or subentire.*

Sa'-lix cor-da'-ta Muhl. **Heart-leaved Willow.** Leaves oblong-lanceolate or narrower, green both sides or scarcely paler beneath; stipules usually large and conspicuous; catkins rather slender.

  A. *Catkins in earliest spring before the leaves.*

Sa'-lix dis'-col-or Muhl. **Glaucous Willow.** Catkins closely sessile, thick, oblong-cylindrical; leaves obovate or elliptic-lanceolate; irregularly crenate-serrate, bright green above, glaucous beneath.

XIII. Order **BET-U-LA'-CE-Æ. BIRCH FAMILY.** —Trees or shrubs; flowers in catkins, or the fertile sometimes in a head or very short catkins; involucre to the nut none, or foliaceous or sac-like.

Fertile flowers with a foliaceous involucre or bladdery bag a .
Fertile flowers with no calyx and no involucre b .
a. Nut small, subtended by an elongated leafy bractlet.................. *Carpinus.* 1
a. Nut small included in a bladdery or enclosed bag...................... *Ostrya.* 2
a. Nut large, with a leafy coriaceous involucre............................ *Corylus.* 3
  b. Fertile scales thin, 3-lobed, deciduous, stamens 2................ *Betula.* 4
  b. Fertile scales thick, entire, persistent, stamens 4................ *Alnus.* 5

1. **CAR-PI'-NUS.**—Trees (or tall shrubs) with smooth, close gray bark, the involucre-like bractlet open, enlarged and foliacious.

Car-pi'-nus car-o-li-ni-a'-na Walt. **Blue** or **Water Beech: Hornbeam; Iron-wood.** Leaves ovate or oblong, sharply doubly-serrate, soon nearly smooth.

2. Genus **OS'-TRY-A.**—Tree with brownish furrowed bark; bractlets tubular, becoming a closed bladdery bag very much larger than the small nut.

Os'-try-a vir-gin'-i-ca (Mill.) Willd. **Hop-Hornbeam; Leverwood; Iron-wood.**—Leaves oblong-ovate, sharply doubly-serrate, downy beneath.

3. Genus **COR'-Y-LUS.**—Shrubs or small trees; leaves thinnish, doubly-toothed, fertile flowers, several in a scaly bud.

Cor'-y-lus a-mer-i-ca'-na Walt. **Wild Hazel-nut.**—Leaves roundish cordate; involucre open above and foliaceous, below coriaceous and downy.

Cor'y-lus ros-tra'-ta Ait. **Beaked Hazel-Nut.** Leaves ovate or ovate-oblong, somewhat cordate; the involucre prolonged into a narrow tubular beak.

4. Genus **BET'-U-LA.** Sterile catkins sessile, long and drooping, flowers opening with or in advance of the leaves; fertile catkins oblong or cylindrical, peduncled; outer bark usually separable in sheets.

A. *Trees; leaves serrate or obscurely lobed.*

b. *Bark brown or yellow gray, more or less sweet aromatic.*

Bet'-u-la len'-ta L. **Sweet. Black. Cherry** or **Mahogany Birch.** Bark of trunk close, outer layer scarcely laminate; leaves oblong-ovate, sub-cordate; fruiting catkins oblong-cylindrical, the scales with short *divergent* lobes.

Bet'-u-la lu'-te-a Mx. f. **Yellow** or **Gray Birch.** — Bark of trunk yellowish gray or silvery gray, detaching in filmy layers; twigs less aromatic than the preceding; leaves ovate-elliptical, scarcely or not at all cordate; fruiting catkins oblong-ovoid, the scales with narrow, barely spreading lobes.

b. *Bark reddish or brown, becoming in young trees conspicuously very loose and torn, and finally in old trees rough like that of the Black Cherry.*

Bet'-u-la ni'-gra L. **River Birch; Red Birch.** — Leaves rhombic-ovate, acutish at both ends, irregularly doubly serrate or obscurely, 9–16 lobed.

A. *Shrubs, leaves crenate, roundish or cuneate.*

Bet'-u-la pu'-mi-la L. **Low Birch.** — Stems 2–8 ft. high, not glandular; young branches and young leaves beneath mostly soft downy; leaves obovate, roundish or orbicular, $\frac{1}{2}$–1$\frac{1}{4}$ in. long, pale beneath.

5. Genus **AL'-NUS.** — Fertile catkins ovoid or oblong, the scales thick and woody in fruit, persistent; shrubs or small trees.

Al'-nus in-ca'-na (L.) Willd. **Speckled** or **Hoary Alder.** Leaves broadly oval or ovate, roundish at base, whitened and mostly downy beneath.

Al'-nus ru-go'-sa (Ehrh.) Koch. (*A. serrulata* Willd.) **Smooth Alder.** Leaves obovate, acute at base, green both sides.

XIV. Order **FA-GA'-CE-Æ. OAK FAMILY.** Trees or shrubs; sterile flowers in catkins; fertile flowers in a cupule of indurate bracts.

Dicotyls or Exogenous Plants.

Sterile flowers in small heads, cupule 4-valved, nuts triangular. ...*Fagus*. 1
Sterile flowers in slender catkins, cupule a pricky bur............*Castanea*. 2
Sterile flowers in slender catkins, cupule scaly, fruit an acorn.....*Quercus*. 3

1. Genus **FA'-GUS.** — Tree with a close smooth ash-gray bark; leaves strongly straight-veined, from slender tapering buds.

Fa'-gus at-ro-pu-nic'-e-a (Marsh.) Sudw. (*F. ferruginea* Ait.) **Beech.** — Leaves oblong-ovate acuminate, distinctly and often coarsely toothed.

2. Genus **CAS-TAN'-E-A.** — Trees or shrubs; sterile flowers in long cylindrical showy white catkins, fertile flowers few in an ovoid and very prickly involucre.

Cas-tan'-e-a den-ta'-ta (Marsh.) Sudw. (*C. satava* var. *americana* Gr.) **Chestnut.** — A very large tree, leaves oblong-lanceolate, long acuminate, serrate with coarse pointed teeth, smooth and green both sides.

Cas-tan'-e-a pu-mi-la (L.) Mill. **Chinquapin.**— A spreading shrub or small tree; leaves oblong, acute, serrate with pointed teeth, *white downy beneath*. Reported in southern Ohio.

3. Genus **QUER'-CUS.** — Mostly large trees; the fertile flowers scattered or somewhat clustered, the ovary enclosed in a scaly bud-like involucre which becomes a *cup* around the acorn.

A. *White Oaks. Leaves lyrate or sinuate pinnatifid.*

Quer'-cus al'-ba L. **White Oak.** — Leaves pale or glaucous beneath, bright green above, obovate-oblong, obliquely cut into 3-9 oblong or linear or obtuse, mostly entire lobes; cup saucer-shaped, rough and naked, much shorter than the ovoid or oblong acorn (1 inch long.)

Quer'-cus mi-nor (Marsh.) Sarg. (*Q. stellata* Wang.) **Iron Oak: Post Oak.** — Leaves grayish or yellowish-downy beneath, *pale and rough above*, *thickish*, sinuately cut into 5-7 rounded, divergent lobes, the upper ones much larger and often truncate or 1-3 notched; cup deep saucer-shaped, naked, one-third or one-half the length of the ovoid acorn ($\frac{1}{2}$-$\frac{3}{4}$ inch long.)

Quer'-cus ma-cro-car'-pa Mx. **Bur Oak.** — Leaves obovate-oblong, lyrate pinnatifid or deeply sinuate lobed or nearly parted, sometimes nearly entire, downy or pale beneath, the lobes sparingly or obtusely toothed or the smaller ones entire; acorn broadly ovoid, 1-1$\frac{1}{2}$ inches long, half immersed or enclosed by the fringed cup.

Spring Flora of Ohio.

A. *Chestnut Oaks. Leaves coarsely sinuate-toothed — or slightly lobed or rather pinnatifid in the swamp White Oak.*

Quer'-cus plat-a-noi'-des (Lam.) Sudw. (*Q. bicolor* Willd.) **Swamp White Oak.** Leaves obovate or oblong-obovate, wedge-shaped at base, coarsely sinuate-crenate, or often rather pinnatifid, usually soft downy and white hoary beneath; fruiting peduncle much longer than the petiole; acorn scarcely 1 inch long, cup fringed or the upper scales pointed.

Quer'-cus pri'-nus L. **Chestnut Oak.** Leaves thick, obovate or oblong to lanceolate, undulately crenate-toothed, pale and minutely downy beneath; fruiting peduncle shorter than the petioles; cup ½ inch wide, mostly tuberculate with hard and stout scales; *acorn large*, sometimes 1-1¼ inches long.

Quer'-cus muh-len-ber'-gi-i Englm. **Yellow Oak.** Leaves slender-petioled, often oblong or even lanceolate, usually acute or pointed, mostly obtuse at base, almost equally and rather sharply toothed; cup shallow, thin, of small appressed scales, 5-7 lines broad; acorn globose or obovate, 7-9 lines long, sweet, hence the common name of the species *Sweet Oak*.

Quer'-cus pri-noi'-des Willd. **Scrub Yellow Oak.** Like the last, but low, usually 2-4 feet high, with smaller and more undulate leaves. Reported for southern Ohio.

A. *Black Oaks. Wood porous and brittle, acorns maturing the second year.*

   b. *Leaves pinnatifid or lobed, not coriaceous.*

      c. *Mature leaves glabrous both sides or nearly so.*

Quer'-cus ru'-bra L. **Red Oak.** Cup saucer-shaped or flat, ¾-1 in. in diameter, acorn oblong-ovoid or ellipsoidal, 1 in. or less in length; leaves moderately (rarely deeply) pinnatifid.

Quer'-cus coc-cin'-e-a Wang. **Scarlet Oak.** — Cup top-shaped or hemispherical with a conical base, 7-9 lines broad, coarsely scaly, covering half or more of the broadly or globular-ovoid acorn, the scales mostly appressed and glabrate, acorn ½-¾ in. long; leaves deeply pinnatifid, the lobes divergent and sparingly cut-toothed; bark of trunk gray, the interior reddish.

Quer'-cus ve-lu-ti'-na Lam. (*Q. coccinea* var. *tinctoria* Gr.) **Black Oak; Quercitron Oak.**— Somewhat resembling the preceding, but the leaves with broader undivided lobes, commonly paler and somewhat

Dicotyls or Exogenous Plants.

pubescent beneath; cup-scales large and loosely imbricated or squarrose when dry, pubescent; bark of trunk darker colored and rougher, thicker and internally orange.

Quer'-cus pa-lus'-tris Du Roi. **Pin Oak: Swamp Spanish Oak.** — Cup flat, saucer-shaped, fine-scaled, 5-7 lines broad, very much shorter than the usually globose or depressed acorn which is 5-7 lines long; leaves deeply pinnatifid with divergent lobes. In low grounds.

  c. *Mature leaves soft-downy beneath.*

Quer'-cus il-i-ci-fo'-li-a Wang. **Bear Oak; Black Scrub Oak.** — Dwarf and straggling, 3-8 ft. high; leaves 2-4 in. long, thickish, obovate, cuneate at base, 3 to 7-lobed, white-downy beneath; acorn ovoid, globular, 5-6 lines long. This species may occur in southern Ohio.

  b. *Leaves entire or with a few teeth or somewhat lobed at the summit, coriaceous.*

Quer'-cus ni'-gra L. **Black Jack; Barren Oak.** — Leaves broadly wedge-shaped, sometimes rounded at base, *widely dilated* and usually somewhat 3-lobed at the summit, rusty pubescent beneath, 4-9 in. long; cup top-shaped.

Quer'-cus im-bri-ca'-ri-a Mx. **Laurel Oak; Shingle Oak.** — Leaves entire, lanceolate-oblong, shining above, downy beneath. A hybrid between this species and Q. coccinea, often called *Quer'cus lea'na* Nutt. (Lea's Oak), occurs sparingly in southern Ohio; its leaves are usually somewhat irregularly and often bristly lobed.

XV. Order **UL-MA'-CE-Æ. ELM FAMILY.** — Trees; flowers polygamous or diœcious-polygamous, not in catkins; fruit a winged samara or a drupe.

Flowers preceding the leaves, fruit winged............ ....*Ulmus*. 1
Flowers appearing with the leaves, fruit a drupe.......... ....*Celtis*. 2

1. Genus **UL'-MUS.** — Calyx 4-5-cleft, fruit a samara, winged all around; leaves strongly straight-veined, oblique at base.

Ul'-mus pu-bes-'cens Walt. (*U. fulva* Mx.) **Red Elm ; Slippery Elm.** — Buds before expansion soft downy with rusty hairs; leaves very rough above, soft downy or slightly rough beneath; branchles downy; inner bark mucilaginous.

Ul'-mus a-mer-i-ca'-na L. **White Elm: American Elm.** — Buds and branchlets glabrous; leaves 2-4 in. long, soft pubescent beneath or soon glabrous; *flowers in close fascicles.*

Ul'-mus ra-ce-mo'-sa Thomas. **Rock Elm: Cork Elm.** — Bud scales downy ciliate, these and the young branchlets somewhat pubescent; branches often with corky ridges; leaves as in the last or the veins more simple and straight; *flowers racemed.*

2. Genus **CEL'-TIS**. — Calyx 5 to 6-parted, persistent; fruit a globular drupe; leaves 3-nerved at base and somewhat oblique; bark of trunk ridged, the furrows discontinuous.

Cel'-tis oc-ci-den-ta'-lis L. **Hackberry: Sugarberry: Nettleberry.** — Leaves reticulated, ovate or ovate-lanceolate, sharply serrate.

XVI. Order **MO-RA'-CE-Æ. MULBERRY FAMILY.** — Flowers racemose, spicate or capitate; calyx becoming fleshy in fruit; trees or large shrubs, with milky juice.

    Leaves serrate, sometimes lobed, 3-nerved ....*Morus.* 1
    Leaves entire, shining, feather-veined........................ ...*Toxylon.* 2

1. Genus **MO'-RUS.** — Flowers in catkin-like spikes, calyx of the fertile spikes succulent, making a juicy, berry-like, oblong fruit; trees.

Mo'r-us ru'-bra L. **Red Mulberry.** — Leaves rough above, downy beneath, those of the young shoots often lobed; fruit dark-purple.

Mo'-rus al'-ba L. **White Mulberry.** — Leaves smooth and shining, sometimes lobed; fruit whitish. Occasionally adventive.

2. Genus **TOX'-Y-LON.** (*Maclura.*) — Staminate flowers in loose short racemes, the pistillate in dense globose heads; trees (or tall shrubs) with milky juice and entire shining leaves.

Tox'-y-lon po-mif'-er-um Raf. (*Maclura aurantiaca* Nutt.) **Osage Orange.** — Cultivated for hedges, rarely escaped.

XVII. Order **LO-RAN-THA'-CE-Æ. MISTLETOE FAMILY.** — Shrubby plants with coriaceous, greenish foliage; parasitic on trees; fruit a globose berry.

Dicotyls or Exogenous Plants.   37

1. Genus **PHO-RA-DEN′-DRON.** — Yellowish-green woody parasites on various deciduous trees; leaves thick, firm, persistent; berry globose, white.

Pho-ra-den′-dron fla-ves′-cens (Ph.) Nutt.   **False Mistletoe.** — In southern Ohio.

XVIII. Order **SAN-TA-LA′-CE-Æ.   SANDALWOOD FAMILY.** — Herbs with entire leaves; calyx-tube adherent with the 1-celled ovary.

1. Genus **CO-MAN′-DRA.** — Calyx bell-shaped or urn-shaped; low smooth perennial herbaceous plants with greenish white flowers.

Co-man′-dra um-bel-la′-ta (L.) Nutt.   **False Toad-Flax.** — Leaves oblong, 1 in. long; fruit globular urn-shaped.

XIX. Order **A-RIS-TO-LO-CHI-A′-CE-Æ.   BIRTHWORT FAMILY.** — Herbs; flowers perfect, apetalous, the calyx conspicuous, cohering with the 6-celled ovary.

1. Genus **AS′-A-RUM.** — Calyx regular, bell-shaped, the limb 3-cleft or parted; rootstocks creeping, aromatic.

As′-a-rum can-a-den′-se L.   **Wild Ginger.** — Soft pubescent; a single pair of large kidney-shaped leaves; calyx brown-purple inside.

XX. Order **PO-LYG-O-NA′-CE-Æ.   BUCKWHEAT FAMILY.** — Herbs; flowers apetalous; stipules in the form of sheaths (called *ochreæ*); joints of the stem swollen.

1. Genus **RU′-MEX.** — The 3 inner sepals enlarged after flowering (in fruit called *valves*) veiny, often with a grain-like tubercle on the back; flowers small, mostly green, in panicled racemes.

Ru′-mex a-ce-to-sel′-la.   Field or **Sheep Sorrel.** — Low, 6-12 in. high; leaves narrow lanceolate or linear, the lowest halbred-form.

XXI. Order **Al-ZOI-DA′-CE-Æ.   INDIAN CHICKWEED FAMILY.** — Leaves opposite, exstipulate; petals wanting; low herbs.

1. Genus **MOL-LU′-GO.** — Sepals 5, white inside, petals none, stamens 5, stigmas 3, capsule 3-celled; prostrate plants.

Mol-lu'-go ver-ti-cil-la'-ta L. **Indian Chickweed: Carpet Weed.**—Leaves spatulate, clustered in whorls at the joints.

XXII. Order **POR-TU-LA-CA'-CE-Æ. PURSLANE FAMILY.**—Herbs; leaves more or less succulent, entire; flowers regular, sepals usually fewer than the 5 petals or sometimes none.

Stamens 5, plants upright, not fleshy........................*Claytonia.* 1
Stamens 7-20, plants fleshy...................................*Portulaca.* 2

1. Genus **CLAY-TO'-NI-A.**—Sepals 2, distinct, persistent, stamens 5, style 3-cleft; simple stems from a deep tuber.

Clay-to'-ni-a vir-gin'-i-ca L. **Spring Beauty.**—Leaves linear-lanceolate, elongated, 3-6 in. long.

Clay-to'-ni-a car-o-li-ni-a'-na Mx. **Broad-leaved Spring Beauty.**—Leaves spatulate oblong, or oval-lanceolate, 1-2 in. long.

2. Genus **POR-TU-LA'-CA.**—Calyx 2-cleft, the tube cohering with the ovary below; style mostly 3-8 parted, stems prostrate, fleshy.

Por-tu-la'ca o-le-ra'-ce-a L. **Purslane.**—Leaves obovate or wedge-shaped; beginning to bloom in June.

XXIII. Order **CAR-Y-O-PHYL-LA'-CE-Æ. PINK FAMILY.**—Herbs, the stems usually swollen at the joints; leaves opposite, entire; stamens distinct.

Flowers with petals a .
Flowers apetalous e .
a. Sepals distinct or nearly so b .
a. Sepals united into a tube or cup...........................*Silene.* 1
   b. Stipules none c .
   b. Stipules present d .
c. Petals 2-cleft or none, styles mostly 3 rarely 4 or 5 pod short. ....*Alsine.* 2
c. Petals notched or 2-cleft, styles 5, pod cylindrical.................*Cerastium.* 3
c. Petals entire, styles usually 3, pod short... ....*Arenaria.* 4
   d. Styles 5, pod 5-valved, leaves whorled......................*Spergula.* 5
   d. Styles 3, pod 3-valved, leaves opposite....... ....*Tissa.* 6
e. Styles usually 4, pod many-seeded c, above .
e. Stigmas 2, sessile, pod 1-seeded..............................*Anychia.* 7

1. Genus **SI-LE'-NE.**—Calyx 5-toothed, styles 3, rarely 4, petals crowned with a scale at the base of the blade.

Dicotyls or Exogenous Plants.

Si-le'-ne car-o-li-ni-a'-na Walt. (*S. pennsylvanica* Mx.) **Wild Pink.**—Stems low (4-8 in.); petals wedge-form, slightly notched and eroded, pink.

Si-le'-ne vir-gin'-i-ca L. **Fire Pink; Catch Fly.** Stems slender, 1-2 ft. high; petals oblong, 2-cleft, deep crimson.

2. Genus **AL-SI'-NE.** (*Stellaria.*) — Sepals 4-5, petals 4-5, stamens 8, 10, or fewer, styles 3, rarely 4 or 5.

   A. *Stems spreading, flaccid, with 1 or 2 pubescent lines.*

Al-si'-ne me'-di-a L. (*Stellaria media* Sm.) **Common Chickweed.** — Lower leaves on hairy petioles; petals shorter than the calyx, stamens 3-10.

Al-si'-ne pu'-be-ra Mx. **Great Chickweed.**—Leaves all sessile; petals longer than the calyx, stamens 10.

   A. *Stems erect or spreading, wholly glabrous.*

     (b. *Petals 2-parted, equalling or surpassing the calyx, bracts scaly.*

Al-si'-ne lon-gi-fo'-li-a (Muhl.) Britt. (*Stellaria longifolia* Muhl.) **Long-leaved Stitchwort.** Leaves linear, acutish at both ends, spreading; the slender pedicels spreading; seeds smooth.

Al-si'-ne lon'-gi-pes (Goldie) Coville. (*Stellaria longipes* Goldie.) **Long-stalked Chickweed.** Leaves ascending, lanceolate or linear-lanceolate, *broadest at the base;* the long pedicels strictly erect, seeds smooth.

Al-si'-ne gra-min'-e-a (L.) Britt. (*Stellaria graminea* L.) **European Chickweed.** — Resembling the last, but the leaves linear-lanceolate, *broadest above the base;* the pedicels widely spreading; seeds strongly rugose.

     (b. *Petals 2-parted, small or none, bracts foliaceous.*

Al-si'-ne bo-re-a'-lis (Bigel.) Britt. (*Stellaria borealis* Bigel.) **Northern Chickweed.** Leaves broadly lanceolate to ovate-oblong. Reported for Northern Ohio.

3. Genus **CE-RAS'-TI-UM.** — Sepals 5, rarely 4, petals 2-lobed or cleft, styles equal to the number of sepals; pod usually elongated, opening at the apex by as many teeth as there are styles.

   A. *Leaves ovate or obovate or oblong-spatulate.*

Ce-ras'-ti-um vis-co'-sum L. **Mouse-ear Chickweed.** Nearly erect, 4-9 in. high, flowers small; pedicels in fruit not longer than the acute sepals; petals shorter than the calyx.

A. *Leaves linear, lanceolate or oblong.*

Ce-ras'-ti-um vul-ga'-tum L. **Larger Mouse-ear Chickweed.** Stems spreading; leaves oblong; flowers larger (sepals 2-3 lines long) the earlier fruiting pedicels much longer than the obtuse sepals; petals equalling the calyx.

Ce-ras'-ti-um lon'-gi-pe-dun-cu-la'-tum Muhl. (*C. nutans* Raf.) **Nodding Mouse-ear Chickweed.** — Stems slender, erect, grooved, diffusely branched, 6-20 in. high; leaves oblong-lanceolate; petals longer than the calyx; pods nodding on the stalks, curved upward, thrice the length of the calyx.

Ce-ras'-ti-um ar-ven'-se L. **Field Chickweed.** — Stems ascending or erect, tufted, 4-8 in. high, naked and few-flowered at the summit; leaves linear or narrowly lanceolate, petals obcordate, more than twice the length of the calyx; pods scarcely longer than the calyx.

Ce-ras-'ti-um ar-ven'-se ob-lon-gi-fo'-li-um (Torr.) Britt. **Mouse-ear Chickweed.** Like the last, but usually taller, leaves narrowly or broadly oblong or oblong-lanceolate, and pod about twice longer than the calyx.

4. Genus **AR-E-NA'-RI-A.** Sepals 5 (rarely 4); petals 5 (rarely 4), entire or barely notched, rarely wanting; low plants with sessile leaves.

Ar-e-na'-ri-a ser-pyl-li-fo'-li-a L. **Thyme-leaved Sandwort.** Leaves ovate, acute, small; plant 2-6 in. high.

Ar-e-na'-ri-a lat-er-i-flo'-ra L. **Sandwort.** Parts of the flower sometimes in fours; leaves oval or oblong, ½-1 in. long; plant erect, branched.

5. Genus **SPER'-GU-LA.** — Sepals, petals and styles each 5, stamens 5 or 10; pod 5-valved, leaves in whorls.

Sper'-gu-la ar-ven'-sis L. **Corn Spurry.** — Leaves numerous, thread-shaped, 1-2 in. long, whorled.

6. Genus **TIS'-SA.** (*Buda*.) — Sepals 5, petals 5, entire, stamens 2-10, styles and valves of pod 3 (rarely 5); leaves linear, opposite.

Dicotyls or Exogenous Plants. 41

Tis'-sa ru'-bra (L.) Britt. (*Buda rubra* Dumor.) **Sand Spurry.** — Corolla small (1½ lines long), pink-red, scarcely equalling or exceeding the calyx.

7. Genus **A-NYCH'-I-A.** Sepals 5, petals wanting, stamens 2-3, rarely 5, stigmas 2, sessile; plants small, many times forked.

A-nych'-i-a di-chot'-o-ma Mx. **Forked Chickweed.** — More or less pubescent, short jointed, low and spreading; flowers nearly sessile.

A-nych'-i-a can-a-den'-sis (L.) B. S. P. (*A. capillacea* DC.) **Forked Chickweed.** — Smooth, longer jointed, slender and erect; leaves broader and longer, 5-15 lines long; flowers more stalked.

XXIV. Order **NYM-PHÆ-A'-CE-Æ. WATER-LILY FAMILY.** — Aquatic, with horizontal perennial rootstocks, leaves peltate or cordate, flowers solitary, large.

Petals adnate to the ovary, large.................................*Castalia*. 1
Petals very small and stamen-like................................*Nymphæa*. 2

1. Genus **CAS-TA'-LI-A.** (*Nymphæa*.) — Sepals 4, green outside; petals numerous, the innermost gradually passing into stamens.

Cas-ta'-li-a o-do-ra'-ta (Dry.) Wv. & Wd. (*Nymphæa odorata* Ait.) **White Water-Lily.** — Leaves orbicular, cordate-cleft at the base to the petiole; flower white, very sweet scented.

2. Genus **NYM-PHÆ'-A.** (*Nuphar*.) — Sepals 5 or 6 or more, colored or partly green outside, roundish concave; petals numerous, scale-like or stamen-like.

Nym-phæ'-a ad'-ve-na Soland. (*Nuphar advena* Ait. f.) **Yellow Pond-Lily; Spatter Dock.** — Sepals 6, unequal; leaves roundish to ovate or almost oblong.

XXV. Order **MAG-NO-LI-A'-CE-Æ. MAGNOLIA FAMILY.** — Trees; the leaf buds covered by stipules; the flowers large, more or less tulip-like.

Leaves oblong, pointed.................................*Magnolia*. 1
Leaves lobed, truncate or notched.......................*Liriodendron*. 2

1. Genus **MAG-NO'-LI-A.** — Tree; buds conical; pistils cohering, forming a fleshy and rather woody cone-like red fruit.

Mag-no'-li-a a-cu-mi-na'-ta L. **Cucumber Tree.** Flower oblong bell-shaped, glaucous-green tinged with yellow; fruit 2-3 in. long; leaves oblong.

2. Genus **LIR-I-O-DEN'-DRON.** Tree, buds flat, pistils flat and scale-form, long and narrow, cohering into an elongated dry cone.

Lir-i-o-den'-dron tu-lip-if'-er-a L. **Tulip Tree; Yellow Poplar; White Wood.** Corolla bell-shaped; leaves very smooth, appearing as if cut off abruptly by a broad shallow notch.

XXVI. Order **A-NO-NA'-CE-Æ. CUSTARD-APPLE FAMILY.** Trees or shrubs; sepals 3, petals 6, filaments short; leaves large, alternate.

1. Genus **A-SIM'-I-NA.** Petals 6, dull purple; stamens in a globular mass; the flowers solitary from the axils of last year's leaves.

A-sim'-i-na tril'-o-ba (L.) Dunal. **Papaw.** Leaves thin, obovate-lanceolate, pointed; fruit oblong, pulpy, edible.

XXVII. Order **RA-NUN-CU-LA'-CE-Æ. CROWFOOT FAMILY.** Plants with sepals, stamens and pistils free (except in one species), petals sometimes absent; sepals often petalloid.

Leaves alternate or radical, the upper often opposite or whorled a .
Leaves all opposite, sepals 4. styles elongated in fruit h .
a. Leaves 2, 5-7-lobed; sepals 3, falling when the solitary flower
  opens..*Hydrastis*. 1
a. Leaves round and cordate or reniform; flowers many, yellow, petals
  none ....*Caltha*. 2
a. Leaves 5-7-parted; flower 1, large, yellowish or whitish, sepals 5-15..*Trollius*. 3
a. Leaves 5-11-lobed, flowers corymbose, white i .
a. Not as above b .
 b. Petals with a spot, pit or scale at the base k .
 b. Petals not as above or wanting c .
  c. Leaves alternate, 2-3-ternately compound, leaflets 2-3-lobed, flowers axillary and terminal; no involucre; root tuber-bearing ....*Isopyrum*. 4
  c. Leaves trifoliate, radical, evergreen, root bright yellow................*Coptis*. 5
  c. Not as above d .
   d. Flowers in a single short raceme, sepals caducous, filaments white, petals 4-10, small, spatulate leaves 2-3 ternate............*Actæa*. 6
   d. Flowers in long racemes, sepals caducous, filaments white, leaves 2-3-ternate, petals 4-8, small, on claws, 2-horned at apex...*Cimicifuga*. 7
   d. Not as above e .
    e. Flowers regular, petals 5, with long spurs, sepals colored..........*Aquilegia*. 8
    e. Sepals petal-like, the upper one spurred; petals 4 rarely 2, the upper pair with long spurs enclosed in the spur of the sepal.....*Delphinium*. 9

Dicotyls or Exogenous Plants. 43

e. Sepals petal-like, the larger upper one hooded and covering the 2 long-
clawed small petals... *Aconitum*. 10
e. Not as above f .
   f. All but the lower leaves opposite or whorled g .
   f. Leaves alternate i .
g. Involucre leaf-like, remote from the flower, peduncles 1-flowered....*Anemone*. 11
g. Involucre close to the flower, calyx-like, leaves simple, 3-lobed......*Hepatica*. 12
g. Involucre compound, at the base of an umbel of flowers..........*Syndesmon*. 13
   h. Sepals 4, leathery, leaves pinnate; woody vine.................*Clematis*. 14
i. Leaves palmately-lobed, the lobes cut and toothed, flowers corymbose ....*Trautvetteria*. 15
i. Leaves, etc., not as above 1 .
   k. Petals 5, yellow, each with a scale at base..................*Ranunculus*. 16
   k. Petals 5, white, a spot or naked pit at base..................*Batrachium*. 17
   k. Petals 8-9, yellow; sepals 3..*Ficaria*. 18
l. Flowers panicled, leaves decompound, petioles dilated at base....*Thalictrum*. 19
l. Flowers in a leafy involucre, leaves finely divided......................*Nigella*. 20

1. Genus **HY-DRAS'-TIS.**— Sepals 3, petal-like, falling off when the flower opens in early spring; a single radical leaf and simple stem with two leaves.

Hy-dras'-tis can-a-den'-sis L. **Golden Seal.**—Ovaries becoming a head of crimson berries.

2. Genus **CAL'-THA.**—Sepals 6-9, petalloid, petals none; leaves round, glabrous; stems hollow, furrowed.

Cal'-tha pa-lus'-tris L. **Marsh Marigold.**— In swamps and wet meadows.

3. Genus **TROL'-LI-US.**—Sepals 5-15, petals 1-lipped, small, numerous; leaves palmately parted; flowar solitary, large.

Trol'-li-us lax'-us Salisb. **Globe Flower.**—In deep swamps.

4. Genus **I-SO-PY'-RUM.**—Sepals 5, petal-like; leaves alternate, 2 to 3-ternately compound, leaflets 2 to 3-lobed.

I-so-py'-rum bi-ter-na'-tum (Raf.) T. & G.— Root thickened here and there into little tubers.

5. Genus **COP'-TIS.** — Sepals 5-7, petal-like; scape 1-flowered; plant low, smooth, perennial; leaves evergreen, shining, leaflets 3.

Cop'-tis tri-fo'-li-a Salisb. **Goldthread.**—Root of long, yellow, bitter fibres.

6. Genus **AC-TÆ´-A.** Sepals 4–5, falling off when the flower expands; leaves 2 to 3-ternately compound, leaflets cleft and toothed; fruit a many-seeded berry.

Ac-tæ´-a ru´-bra (Ait.) Willd. (*A. spicata* var. *rubra* Ait.) **Red Baneberry.**—Raceme ovate, pedicels slender; berries cherry-red (or sometimes white), oval; petals rhombic, spatulate, much shorter than the stamens.

Ac-tæ´-a al´-ba (L.) Mill. **White Baneberry.**—Raceme oblong, pedicels thickened in fruit, red; berries white, globular oval; petals slender, mostly truncate.

7. Genus **CIM-I-CIF´-U-GA.**— Sepals 4 or 5, falling off as soon as the flower expands; flowers in elongated racemes; leaves 2 to 3-ternately divided.

Cim-i-cif´-u-ga ra-ce-mo´-sa (L.) Nutt. **Black Snakeroot; Black Cohosh.** —Stem 3–8 ft. high; racemes in fruit 1–3 ft. long.

8. Genus **AQ-UI-LE´-GI-A.**—Sepals 5, colored like the spurred petals; leaves 2 to 3-ternately compound; flowers large and showy.

Aq-ui-le´-gi-a can-a-den´-sis L. **Wild Columbine.**—Spurs nearly straight.

Aq-ui-le´-gi-a vul-ga´-ris L. **Garden Columbine.**—Spurs hooked. Sometimes escaped from cultivation.

9. Genus **DEL-PHIN´-I-UM.**— The upper sepal prolonged into a spur and enclosing the spurs of two petals; leaves palmately divided or cut; flowers in terminal racemes.

Del-phin´-i-um tri-cor´-ne Mx. **Dwarf Larkspur.** — Pistils 3; raceme few-flowered, loose; flowers bright blue, sometimes white; pods strongly diverging.

Del-phin´-i-um car-o-li-ni-a´-num Walt. (*D. azureum* Mx.) **Larkspur.** —Pistils 3; raceme strict; flowers sky-blue or whitish; pods erect.

Del-phin´-i-um con-sol´-i-da L. **Field Larkspur.**—Pistil single, petals 2, united into one body; leaves dissected into narrow linear lobes. Introduced and flowering later.

10. Genus **AC-O-NI´-TUM.**— The upper (larger) sepal hooded, concealing two small spur-shaped, clawed petals; leaves palmately cleft; flowers blue, showy, in racemes.

Dicotyls or Exogenous Plants.

Ac-o-ni'-tum nov-e-bor-a-cen'-se Gr. **Aconite: Monkshood; Wolfsbane.**
Plant 2 ft. high, erect, leafy, the summit and the loosely flowered raceme pubescent; the broadly-cuneate divisions of the leaf 3-cleft and incised.

Ac-o-ni'-tum un-ci-na'-tum L. **Aconite: Monkshood; Wolfsbane.** — Plant glabrous, erect but weak and disposed to climb; the lobes of the leaves ovate-lanceolate, coarsely-toothed.

11. Genus **AN-E-MO'-NE.** — Sepals petal-like; leaves radical, those of the stem 2 or 3, opposite or whorled, forming an involucre remote from the flower; peduncles 1-flowered; achenes flattened, not ribbed.

An-e-mo'-ne vir-gin-i-a'-na L. **Anemone: Windflower.** — Plant 2-3 ft. high; involucral leaves 3, 3-parted, long petioled; head of fruit oval or oblong.

An-e-mo'-ne can-a-den'-sis L. (*A. pennsylvanica* L.) **Anemone: Windflower.** — Plant 1-1½ ft. high, involucre sessile, 3-leaved, the naked peduncle with a pair of branches and a 2-leaved involucre.

An-e-mo'-ne quin-que-fo'-li-a L. (*A. nemorosa* var. *quinquefolia* Gr.) — Plant low, stem simple from a filiform rootstock; involucre of 3, long-petioled leaves; leaflets 5, wedge-shaped or oblong, toothed or cut.

12. Genus **HE-PAT'-I-CA.** — Involucre resembling a calyx; leaves all radical, cordate and 3-lobed, persistent through the winter.

He-pat'-i-ca he-pat'-i-ca (L.) Karst. (*H. triloba* Chaix.) **Liver-leaf; Hepatica.** — Lobes of the leaves obtuse or rounded; those of the involucre also obtuse.

He-pat'-i-ca a-cu'-ta (Ph.) Britt. (*H. acutiloba* DC.) **Liverleaf; Hepatica.** — Leaves with 3 ovate and pointed lobes, or sometimes 5-lobed; involucral leaves acute or acutish.

13. Genus **SYN-DES'-MON.** (*Anemonella*). — Flowers in an umbel surrounded by an involucre; achenes terete, 8-10-ribbed.

Syn-des'-mon tha-lic-troi'-des (L.) Hoffing. (*Anemonella thalictroides* Spach.) **Rue Anemone.** Stem and radical leaf from a cluster of thickened tuberous roots; leaves 2-3-ternately compound, leaflets roundish, somewhat 3-lobed, cordate at base; sepals rarely 3-lobed.

14. Genus **CLEM'-A-TIS.** — Perennial woody vine with opposite compound leaves; sepals 4, leathery; achenes numerous, tailed with the long, persistent, hairy styles.

Clem'-a-tis vir-gin-i-a'-na L., **Virgin's-Bower.** Flowers rather small, cymose, paniculate, sepals whitish; leaflets 3.

Clem'-a-tis vi-or'-na L., **Leather Flower: Clematis.**—Sepals purplish, very thick and leathery; leaflets 3-7; flowers large, solitary, on long pedicels.

15. Genus **TRAUT-VET-TE'-RI-A.** — Sepals 3-5, petal-like, very caducous; alternate palmately-lobed leaves and white corymbose flowers.

Traut-vet-te'-ri-a car-o-li-nen'-sis (Walt.) Vail. (*T. palmata* F. & M.) **False Bugbane.**—Growing in moist ground along streamlets.

16. Genus **RA-NUN'-CU-LUS.**—Sepals 5, green; petals 5, with a scale at base; achenes numerous, mostly pointed; flowers solitary or somewhat corymbed.

A. *Aquatic; immersed leaves filiformly dissected.*

Ra-nun'-cu-lus del-phin-i-fo'-li-us Torr. (*R. multifidus* Ph.) **Yellow Water Crowfoot.**—Leaves all repeatedly 3-forked into long filiform divisions; petals 5-8, deep bright yellow.

A. *Terrestrial, but often in wet places.*

  b. *Root-leaves not divided to the very base.*

Ra-nun'-cu-lus ab-or-ti'-vus L.. **Small-flowered Crowfoot.**—*Plant glabrous*; the primary root-leaves round-cordate or reniform, barely crenate, the succeeding ones often 3-lobed or 3-parted; petals pale yellow, shorter than the reflexed calyx; carpels in globose heads.

Ra-nun'-cu-lus mi-cran'-thus Gr. (*R. abortivus* var. *micranthus* Gr.) **Small-flowered Crowfoot.**—Plant pubescent, otherwise much like the preceding, but root-leaves seldom at all heart-shaped.

Ra-nun'-cu-lus scel-er-a'-tus L. **Cursed Crowfoot.**—Glabrous, root-leaves 3-lobed; petals scarcely exceeding the calyx; head of carpels oblong or cylindrical; stem thick and hollow.

  b. *Leaves variously cleft or divided.*
    c. *Achenes with long recurved beak.*

Dicotyls or Exogenous Plants. 47

Ra-nun'-cu-lus re-cur-va'-tus Poir. **Hooked Crowfoot.**—Hirsute; leaves long-petioled, deeply 3-cleft, large; petals shorter than the reflexed calyx, pale.

c. *Styles long and attenuate, stigmatose at the tip.*

Ra-nun'-cu-lus fas-cic-u-la'-ris Muhl. **Early Crowfoot.**—Pubescent with close-pressed silky hairs; lateral divisions of the radical leaves *sessile;* carpels scarcely margined.

Ra-nun'-cu-lus sep-ten-tri-o-na'-lis Poir. **Northern Crowfoot.**—Hairy or nearly glabrous; leaves 3-divided, the divisions usually stalked; carpels strongly margined.

c. *Style subulate, stigmatose along the inner margin.*

Ra-nun'-cu-lus re'-pens L. **Creeping Crowfoot.**— Closely resembling the last species; leaves often white-variegated or spotted; commencing to flower somewhat later.

Ra-nun'-cu-lus penn-syl-van'-i-cus L. f. **Bristly Crowfoot.** Stout, erect; hirsute with widely spreading bristly hairs; leafy to the top; 2 ft. high; flowers inconspicuous; head of carpels oblong.

Ra-nun'-cu-lus his'-pi-dus Mx. **Hispid Crowfoot.**—Resembling the last but the ascending or reclining stems few-leaved, not always hirsute; head of carpels globose or oval. A northern species reported for Ohio.

Ra-nun'-cu-lus bul-bo'-sus L. **Buttercups: Bulbous Crowfoot.** — Hairy; stem erect, 1 ft. high, from a bulbous base; radical leaves 3-divided, the lateral divisions sessile; peduncles furrowed.

Ra-nun'-cu-lus ac'-ris L. **Buttercups; Tall Crowfoot.** — Hairy, stem erect, 2-3 ft. high; leaves 3-divided, the divisions all sessile; peduncles not furrowed.

17. Genus **BA-TRA'-CHI-UM.** — Sepals 5, the 5 petals each with a spot or naked pit at base, white or only the claw yellow; aquatic or sub-aquatic plants with the immersed leaves repeatedly dissected into filiform divisions.

Ba-tra'-chi-um di-var-i-ca'-tum (Schrk.) Wimm. (*Ranunculus circinatus.* Sibth.) **Stiff White Water Crowfoot.** — Leaves sessile, the divisions short, spreading in one roundish plane, rigid, not collapsing when withdrawn from the water.

Ba-tra'-chi-um tri-cho-phyl'-lum (Chx.) Bossch. (*Ranunculus aqualilis* var. *trichophyllus* Gr.) **Common White Water Crowfoot.**—Leaves mostly petioled, their divisions rather long and soft, collapsing more or less when withdrawn from the water.

18. Genus **FI-CA'-RI-A.**—Sepals 3; petals 8-9, yellow; peduncles scape-like; leaves roundish-cordate, roots tuberous thickened.

Fi-ca'-ri-a fi-ca'-ri-a (L.) Karst. (*Ranunculus ficaria* L.) Occasionally escaped from gardens.

19. Genus **THA-LIC'-TRUM.**—Sepals 4-5, usually caducous, petals none; leaves alternate, 2 to 3-ternately compound, the petioles dilated at base; flowers dioecious.

Tha-lic'-trum di-oi'-cum L. **Early Meadow Rue.**—Stems 1-2 ft. high, smooth and pale or glaucous; leaves all with general petioles.

Tha-lic'-trum pur-pu-ras'-cens L. **Purplish Meadow Rue.**—Stems 2-4 ft. high, usually purplish; stem leaves sessile or nearly so.

20. Genus **NI-GEL'-LA.**—Calyx of 3 colored sepals; petals 5, cleft, the 5 ovaries united below; leaves finely divided.

Ni-gel'-la dam-as-ce'-na L. **Fennel Flower; Ragged Lady.** Flower bluish, surrounded by a finely-divided leafy involucre. Introduced.

XXVIII. Order **BER-BER-I-DA'-CE-Æ. BARBERRY FAMILY.** Shrubs or herbs; petals 6-9, stamens 6-18; anthers usually opening by 2 valves or lids hinged at the top.

Petals 6-9, stamens 6-18; fruit many-seeded a.
Petals 6, stamens 6; fruit few-seeded b.
a. Petals 6-9, stamens 12-18, anthers not opening by valves..........*Podophyllum*. 1
a. Petals and stamens usually 8, anthers opening by valves..........*Jeffersonia*. 2
b. Herbs with greenish flowers, petals thick and short........*Caulophyllum*. 3
b. Shrubs with yellow flowers and wood, petals with 2 glands.....*Berberis*. 4

1. Genus **POD-O-PHYL'-LUM.**—Creeping rootstock; stem 2-leaved, 1-flowered; sepals 6, fugacious; fruit a large fleshy berry.

Pod-o-phyl'-lum pel-ta'-tum L. **May Apple; Mandrake.**—Leaves peltate, 5 to 9-parted; flower white, large.

2. Genus **JEF-FER-SO'-NI-A.**—Sepals usually 4, fugacious; pod pear-shaped, opening by a lid; leaves radical, with 2 half-ovate leaflets.

Dicotyls or Exogenous Plants.

Jef-fer-so′-ni-a di-phyl′-la (L.) Pers. **Twin-leaf.** Scapes 1-flowered; flower white, 1 in. broad.

3. Genus **CAU-LO-PHYL′-LUM.** Stem with a tri-ternately compound sessile leaf and terminated by a small raceme or panicle of yellowish-green flowers.

Cau-lo-phyl′-lum tha-lic-troi′-des (L.) Mx. **Blue Cohosh : Pappoose root.** — Flowers appearing while the leaf is yet small.

4. Genus **BER′-BER-IS.** — Shrub; yellow flowers in drooping racemes; berries oblong, scarlet.

Ber′-ber-is vul-ga′-ris L. **Barberry.** Somewhat spiny. Often escaped from cultivation.

XXIX. Order **CAL-Y-CAN-THA′-CE-Æ. CALYCANTHUS FAMILY.** — Shrubs with opposite, entire leaves; the sepals and petals similar and indefinite; the fruit like a rose-hip.

1. Genus **BUETT-NE′-RI-A.** (*Calycanthus*). — Calyx of many sepals united below, lurid purple; petals similar; stamens numerous; leaves opposite, entire; shrubs.

Buett-ne′-ri-a flor′-i-da (L.) Kear. (*Calycanthus floridus* L.) **Sweet-scented Shrub; Carolina Allspice: Calycanthus.** — Leaves oval, soft-downy underneath; cultivated.

Buett-ne′-ri-a fer′-til-is (Walt.) Kear. (*Calycanthus lævigatus* Willd. and *Calycanthus glaucus* Willd.) Leaves oblong, oblong-ovate, or ovate-lanceolate, green and glabrous both sides, or glaucous-white beneath.

XXX. Order **LAU-RA′-CE-Æ. LAUREL FAMILY.** — Aromatic trees or shrubs; leaves alternate, simple; anthers opening by valves; ovary free from the colored calyx, 1-ovuled.

Flowers appearing with the leaves, anthers 4-valved; trees.......*Sassafras.* 1
Flowers appearing before the leaves, anthers 2-valved; shrubs......*Benzoin.* 2

1. Genus **SAS′-SA-FRAS.** Trees with spicy-aromatic bark and very mucilaginous twigs and foliage; flowers in clustered and peduncled corymbed racemes.

Sas'-sa-fras sas'-sa-fras (L.) Karst. (*S. officinale* Nees.) **Sassafras.** — Leaves ovate, entire or some of them 3-lobed.

2. Genus **BEN-ZO'-IN.** — Shrubs, spicy; flowers in umbel-like almost sessile clusters, appearing before the leaves.

Ben-zo'-in ben-zo'-in (L.) Coult. (*Lindera benzoin* Blm.) **Spice-bush; Benjamin-bush; Wild Allspice.** — Leaves oblong-ovate, pale underneath.

XXXI. Order **PA-PAV-ER-A'-CE-Æ. POPPY FAMILY.** — Herbs with milky or colored juice; sepals fugacious, ovary 1-celled; placentas 2 or more, parietal.

Flowers regular a
Flowers irregular c
a. Flowers yellow, petals 4, leaves pinnately parted b .
a. Flowers white, petals 8–12, leaves palmately lobed................*Sanguinaria.* 1
  b. Pod bristly, style distinct, stigmas 3–4.........................*Stylophorum.* 2
  b. Pod linear, smooth, style almost none, stigmas 2............*Chelidonium.* 3
c. Corolla cordate or 2-spurred.................................*Bicuculla.* 4
c. Corolla with but one petal spured................................*Capnoides.* 5

1. Genus **SAN-GUI-NA'-RI-A.** — Sepals 2, petals 8–12; rootstock thick, containing red-orange acrid juice; leaf single, round, palmately lobed.

San-gui-na'-ri-a can-a-den'-sis L. **Blood-root.** — Flower single on a scape, white.

2. Genus **STY-LOPH'-O-RUM.** — Sepals 2, hairy, petals 4; stems naked below, 2-leaved or 1 to 3-leaved above, and 1 — few-flowered at the summit; juice yellow.

Sty-loph'-o-rum di-phyl'-lum (Mx.) Nutt. **Celandine Poppy.** — Leaves pale or glaucous beneath, 5–7-sinuately lobed; flowers deep yellow, 2 in. broad.

3. Genus **CHEL-I-DO'-NI-UM.** — Sepals 2, petals 4; biennial herb, stems brittle, juice saffron-colored; leaves pinnately divided or 2-pinnatifid and toothed or cut.

Chel-i-do'-ni-um ma'-jus L. **Celandine.** — Flowers small, yellow, in a pedunculate umbel; buds nodding.

Dicotyls or Exogenous Plants.

4. Genus **BI-CU-CUL'-LA**. (*Dicentra*).—Petals 4, slightly cohering, forming a heart-shaped or 2-spurred corolla; filaments slightly united in two sets; leaves compound-dissected.

Bi-cu-cul'-la cu-cul-la'-ri-a (L.) Millsp. (*Dicentra cucullaria* DC.) **Dutchman's Breeches.** Bulb granulate; corolla with 2 divergent spurs longer than the pedicel; raceme simple.

Bi-cu-cul'-la can-a-den'-sis (Goldie) Millsp. (*Dicentra canadensis*, DC.) **Squirrel-corn.** Subterranean stems with scattered grain-like tubers; corolla nearly heart-shaped, the spurs very short and rounded; raceme simple.

Bi-cu-cul'-la ex-im'-i-a (Ker.) Millsp. (*Dicentra eximia* DC.) **Ear-Drop.** Subterranean shoots scaly; raceme compound, clustered; flowers oblong, spurs very short.

5. Genus **CAP-NOI'-DES**. (*Corydalis*).—Corolla 1-spurred at the base, deciduous, style persistent; flowers in racemes; plants pale or glaucous; leaves compound.

Cap-noi'-des sem-per-vi'-rens (L.) Borck. (*Corydalis glauca* Ph.) **Pale Corydalis.**—Stems strict; flowers purplish or rose color with yellow tips; racemes panicled.

Cap-noi'des flav'-u-lum (Raf.) Kuntze. (*Corydalis flavula* DC.) **Yellow Corydalis.**—Low, ascending; outer petals wing-crested on the back; corolla pale yellow, 3-4 lines long; spur very short.

Cap-noi'-des au'-re-um (Willd.) Kuntze. (*Corydalis aurea* Willd.) **Golden Corydalis.** Corolla golden yellow, $\frac{1}{2}$ in. long, the slightly decurved spur about one-half as long, shorter than the pedicel; outer petals merely carinate on the back, not crested.

XXXII. Order **CRU-CIF'-ER-Æ. MUSTARD FAMILY.**—Herbs with cruciform flowers, sepals 4, petals 4; stamens 6 (rarely only 2 or 4), 4 long and 2 short; fruit a silique or silicle.

Flowers white, purple, etc., but not yellow a .
Flowers yellow, yellowish or greenish-yellow b .
a. Pod obcordate-triangular, many-seeded d .
a. Pod roundish, flattened, notched at the top, 1 seed in each cell......*Lepidium*. 1
a. Not as above c .
   b. Pod elongated, terete, *close-pressed* to the stem; leaves runcinate.................................*Sisymbrium*. 2

Spring Flora of Ohio.

  b. Pod elongated, somewhat 1-sided, valves nerved; lower leaves lyrate .... *Barbarea.* 3
  b. Not as above e .
 c. Flowers purplish, stem-leaves auricled, pods 1-1½ in. long .... *Iodanthus.* 4
 c. Flowers white or yellow, leaves pinnate, pinnatifid or very large and crenate, pod oblong-linear to globular, terete or nearly so, seeds in two rows in each cell .... *Roripa.* 5
 c. Flowers white or purple, stems leafy, leaves simple or pinnate, pod linear, flattened, seeds in a single row in each cell .... *Cardamine.* 6
 c. Pod, etc., like the last; stem simple, leafless below, bearing 2 or 3 petioled compound leaves above the middle; flowers large, white or purple, in corymbs or short racemes .... *Dentaria.* 7
 c. Not as above d .
  d. Pod obcordate-triangular, flowers small, white .... *Bursa.* 8
  d. Pod oval, oblong, or linear, flat; plants low .... *Draba.* 9
  d. Pod long-linear, flattened; valves 1-nerved or veiny f .
 e. Leaves pinnate, pinnatifid, or very large and crenate c, above .
 e. Leaves 2-pinnatifid, the divisions small and toothed. .... *Descurainia.* 10
 e. Leaves not as above f .
  f. Pod somewhat 1-sided, leaves obovate or oblong, not clasping .... *Stenophragma.* 11
  f. Pod flattened, leaves simple or compound, often clasping at base. *Arabis.* 12

1. Genus **LE-PID'-I-UM.** - Pods roundish, notched at the top, much flattened contrary to the narrow partition, the valves keeled; flowers small, white or greenish.

A. Leaves all with a tapering base.

  Le-pid'-i-um vir-gin'-i-cum L. **Pepper-grass.** Pod marginless or obscurely margined at the top; petals present.

  Le-pid'-i-um in-ter-me'-di-um Gr. **Pepper-grass.** – Pod minutely wing-margined at the top; petals usually minute or wanting.

  Le-pid'-i-um ru-der-a'-le L. **Pepper-grass.** — Pods smaller, oval; petals always wanting; plant more diffuse.

A. Stem-leaves with a sagittate base.

  Le-pid'-i-um cam-pes'-tre (L.) Gr. **Pepper-grass.** — Minutely soft-downy; pods ovate, winged.

2. Genus **SIS-YM'-BRI-UM.** Pods terete, upright and close-pressed to the stem, flowers very small, pale yellow.

  Sis-ym'-bri-um of-fic-i-na'-le (L.) Scop. **Hedge Mustard.**—Very branching, 2–3 feet high; leaves runcinate.

3. Genus **BAR-BA-RE'-A**. Pod linear, terete or somewhat 4-sided, the valves keeled by a nerve; flowers yellow.

Bar-ba-re'-a bar-ba-re'-a (L.) Macm. (*B. vulgaris* R. Br.) **Winter Cress: Yellow Rocket.** Smooth, lower leaves lyrate, the terminal division large and round, flowers yellow.

4. Genus **I-O-DAN'-THUS**. (*Thelypodium*) Pod terete or nearly so, valve 1-nerved, seeds one row in each cell; stout plants, with large purplish flowers.

I-o-dan'-thus pin-na-tif'-i-dus (Mx.) Prantl. (*Thelypodium pinnatifidum* Wats.) **Rocket Cress.**— Root-leaves round or cordate, stem leaves auricled.

5. Genus **ROR'-IP-A**. (*Nasturtium*.) — Pod short, varying from oblong-linear to globular, terete or nearly so, valves nerveless; seeds mostly in two irregular rows; aquatic or marsh plants, with yellow or white flowers.

A. *Petals white, twice the length of calyx; leaves pinnate.*

Ror'-ip-a nas-tur'-ti-um (L.) Rusby. (*Nasturtium officinale* R. Br.) **True Water Cress.**— Perennial, stems spreading and rooting; leaflets 3-11, pods 6-8 lines long. In brooks and ditches.

A. *Petals yellow or yellowish, small.*
b. *Pods longer than the pedicels.*

Ror'-ip-a ses-sil-i-flo'-ra (Nutt.) Hitch. (*Nasturtium sessiliflorum* Nutt.) **Water Cress.** Leaves obtusely incised or toothed, obovate or oblong; flowers minute, nearly sessile; pods elongated-oblong, 5-6 lines long.

Ror'-i-pa ob-tu'-sa (Nutt.) Britt. (*Nasturtium obtusum* Nutt.) **Water Cress.**— Leaves pinnately parted or divided, the divisions roundish and obtusely toothed or repand; pod linear-oblong to short oval.

b. *Pods mostly shorter than the pedicels.*

Ror'-i-pa pa-lus'-tris (L.) Bess. (*Nasturtium palustre* DC.) **Marsh Cress.**—Leaves pinnately cleft or parted or the upper laciniate, the lobes oblong, cut-toothed; pods oblong, ellipsoid.

Ror'-i-pa his'-pi-da (Desv.) Britt. (*Nasturtium palustre* var. *hispidum* Gr.) **Marsh Cress.**— Like the last, but the pods ovoid or globular.

A. *Petals white, leaves undivided, or pinnatifid.*

Ror'-ip-a a-mer-i-ca'-na (Gr.) Britt. (*Nasturtium lacustre* Gr.) **Lake Cress.**—Aquatic; leaves oblong, entire, serrate or pinnatifid, the immersed ones dissected into capillary divisions; pod ovoid, a little longer than the style; flowering later than the preceding species.

Ror'-ip-a ar-mo-ra'-ci-a (L.) Hitch. (*Nasturtium armoracia* Fr.) **Horse-radish.**—Root-leaves very large, oblong, crenate, rarely pinnatifid, those of the stem lanceolate; pods globular (seldom formed.)

6. Genus **CAR-DA-MI'-NE.**—Pod linear, flattened, the partition and placentas thick, seeds in a single row in each cell; flowers white or purple.

A. *Leaves simple.*

Car-da-mi'-ne bul-bo'-sa (Schreb.) B. S. P. (*C. rhomboidea* DC.) **Spring Cress.**—Stems simple, upright, from a tuberous base, the slender rootstock bearing small tubers; root-leaves round, often cordate, lower stem-leaves ovate or rhombic-oblong, the upper sub-lanceolate and sessile; flowers large, white.

Car-da-mi'-ne doug-las'-si (Torr.) Britt. (*C. rhomboidea* var. *purpurea* Torr.) **Purple Spring Cress.**—Low, 4–6 in. high, usually sub-pubescent; flowers rose-purple, appearing very early.

Car-da-mi'-ne ro-tun-di-fo'-li-a Mx. **Mountain Water Cress.** Stems branching, weak or decumbent, making long runners, root fibrous; leaves roundish, angled, often cordate, petioled; flowers white, not large.

A. *Leaves pinnate.*

Car-da-mi'-ne pra-ten'-sis L. **Cuckoo-Flower.** Flowers showy, petals thrice the length of the calyx; leaflets of the lower leaves rounded and stalked, those of the upper oblong or linear. Northward; perhaps incorrectly reported for Ohio.

Car-da-mi'-ne hir-su'-ta L. **Small Bitter Cress.**—Flowers small, white; stems 3–20 in. high, erect or spreading from the spreading cluster of root-leaves; pedicels nearly or quite erect and the pods appressed. This species has been reported for Ohio, but it is very rare according to Prof. Britton, and it is likely that the following has been mistaken for it.

Dicotyls or Exogenous Plants.

Car-da-mi'-ne penn-syl-van'-i-ca Muhl. **Bitter Cress.**—Differs from the last in being large and very leafy, often 2 ft. high; the leaf-segments larger, narrowed rather than rounded at the base as in C. hirsuta, and they have a decided tendency to be decurrent on the rachis, are thinner and more lobed, and the pedicels somewhat spreading.

Car-da-mi'-ne ar-e-nic'-o-la Britt. **Bitter Cress.**—This, much resembling the preceding two species, is much branched at base, the numerous ascending or erect leafy stems 6-12 in. high; leaves nearly erect, divided into linear or linear-oblong segments; fruiting pedicels ascending and the pods strictly erect. Grows in open, sandy, moist soil.

7. Genus **DEN-TA'-RI-A.**— Pod lanceolate, flat, seeds in one row; rootstocks long, horizontal, fleshy, of a pleasant pungent taste; the simple stems with 2 or 3 compound leaves; flowers large, white or purple in a corymb or short raceme.

Den-ta-ri'-a di-phyl'-la L. **Pepper-Root; Toothwort.**— Rootstock long and continuous, toothed; stem-leaves 2, leaflets rhombic-ovate or oblong-ovate.

Den-ta-ri'-a la-cin-i-a'-ta Muhl. **Pepper-Root.** Rootstock tuberous, more or less monoliform, tubers deep-seated; stem-leaves 3-parted, the lateral segments often 2-lobed, all broadly oblong to linear, more or less toothed.

8. Genus **BUR'-SA.**— Pod obcordate-triangular, flattened, flowers small, white; raceme becoming much elongated.

Bur'-sa bur'-sa-pas-to'-ris (L.) Weber. (*Capsella bursa-pastoris* Mœnch.) **Shepherd's Purse.** - Root-leaves clustered, pinnatifid or toothed, stem-leaves sagittate and clasping.

9. Genus **DRA'-BA.** - Pod oval, oblong or even linear, flat; seeds in 2 rows in each cell; low plants with entire or toothed leaves, and white flowers; pubescence often stellate.

Dra'-ba in-ca'-na ar-a-bi'-sans (Mx.) Wats. **Draba.**—Hoary-pubescent, leafy-stemmed; leaves oblanceolate or the cauline ones lanceolate to ovate; pod glabrous, acuminate or acute, twisted, beaked with a distinct style.

Dra'-ba car-o-li-ni-a'-na Walt. **Draba.**—Small, 1-5 in. high; leaves obovate, entire; pods broadly linear, smooth, much longer than the ascending pedicels; raceme short or corymbous in fruit.

Dra'-ba ver'-na L. **Whitlow Grass.**—Leaves all radical, oblong or lanceolate; scapes 1–3 in. high, *petals 2-cleft.*

10. Genus **DES-CU-RAI'-NI-A.**—Pods oblong, club-shaped, or oblong-linear, shorter than the mostly horizontal pedicels; seeds 2-ranked in each cell; leaves 2-pinnatifid, often hoary or downy, the divisions small and toothed.

Des-cu-rai'-ni-a pin-na'-ta (Walt.) Britt. (*Sisymbrium canescens* Nutt.) **Tansy Mustard.**—Not common.

11. Genus **STEN-O-PHRAG'-MA.** Pods linear, somewhat 4-sided, longer than the slender spreading pedicels; leaves obovate or oblong; plant slender, branched.

Sten-o-phrag'-ma tha-li-a'-na (L.) Celak. (*Sisymbrium thaliana* Gaud.) **Mouse-ear Cress.**—Introduced from Europe; not common.

12. Genus **AR'-A-BIS.** Pod linear, flattened, the valves more or less 1-nerved in the middle, or longitudinally veiny; flowers white or purple.

A. *Seeds in one row in each cell, orbicular or nearly so; plants diffuse or erect.*
  b. *Low, diffuse or spreading from the base.*

Ar'-a-bis vir-gin'-i-ca (L.) Trel. (*A. ludoviciana* Meyer.) **Rock Cress.** Leaves pinnately parted into oblong or linear few-toothed or entire divisions; flowers small, white; pods rather broadly linear, spreading, flat. Southern Ohio.

  b. *Stems erect, simple, leafy; pods erect or ascending.*

Ar'-a-bis pa'-tens Sulliv. **Rock-cress.**—Downy with spreading hairs, erect, 1–2 ft. high; stem-leaves oblong, ovate, partly clasping by the cordate base; petals white, 4 lines long, twice the length of the calyx; *pedicels slender, spreading.*

Ar'-a-bis hir-su'-ta (L.) Scop. **Rock-cress.**—Rough-hairy, sometimes smoothish, 1–2 ft. high; stem-leaves oblong or lanceolate, clasping by an arrow-shaped or cordate base; petals greenish-white, small, but longer than the calyx; *pedicels and pods strictly upright.*

  b. *Stems erect, simple, leafy; pods recurved, spreading or pendulous.)*

Ar'-a-bis læv-i-ga'-ta (Muhl.) Poir. **Rock-cress.**—Smooth and glaucous; stem-leaves partly clasping by the arrow-shaped base, lanceolate or linear; pods long and narrow, recurved-spreading.

Dicotyls or Exogenous Plants.

Ar'-a-bis can-a-den'-sis L. **Sickle-pod.**—Stem upright, smooth above; steam-leaves pubescent, pointed at both ends, sessile, the lower toothed; pods very flat, scythe-shaped, hanging on rough, hairy pedicels.

A. *Seeds not so broad as the partition, in two more or less distinct rows in each cell; erect and leafy-stemmed plants.*

Ar'-a-bis per-fo-li-a'-ta Lam. **Tower Mustard.** Tall, 2-4 ft. high, glaucous; stem-leaves oblong or ovate-lanceolate, entire; petals yellowish white, little longer than the calyx.

Ar'-a-bis brach-y-car'-pa (T. & G.) Britt. (*A. confinis* Wats.) **Rock Cress.**—Scarcely glaucous, 1-3 ft. high, pubescence below finely stellate; stem-leaves lanceolate or oblong-linear, 1-2 in. long, entire, with narrow auricles, or the lowest spatulate and toothed; petals white or rose-color, fully twice the length of the calyx.

A. *Seeds oblong or elliptical, very small; plants branching from the base.*

Ar'-a-bis ly-ra'-ta L. **Rock Cress.**—Mostly glabrous except the lyrate pinnatifid root-leaves; stem-leaves spatulate or linear with a tapering base, sparingly toothed or entire; petals white, much longer than the yellowish calyx.

Ar'-a-bis den-ta'-ta T. & G. **Rock Cress.**—Roughish pubescent, slender, 1-2 ft. high; leaves oblong, very obtuse, unequally and sharply toothed, those of the stem numerous, half-clasping and eared at the base.

XXXIII. Order **SAX-I-FRA-GA'-CE-Æ. SAXIFRAGE FAMILY.**—Herbs or shrubs; leaves opposite or alternate; stamens mostly definite and with the petals on the calyx.

    Herbs a
    Shrubs b
a. Ovary 2 rarely 3 -celled, or with 2 or 3 nearly distinct carpels b .
a. Ovary 1-celled, with 2 parietal placentas c .
    b. Petals 5, stamens 5..*Sullivantia*. 1
    b. Petals 5, stamens 10...*Saxifraga*. 2
  c. Petals entire, stamens 10......................................*Tiarella*. 3
  c. Petals small, entire, stamens 5..................................*Heuchera*. 4
  c. Petals small, pinnatifid, stamens 10............................*Mitella*. 5
  c. Petals none, stamens 8-10.................................*Chrysosplenium*. 6
    d. Leaves opposite, fruit a capsule.........................*Hydrangea*. 7
    d. Leaves alternate, fruit a berry..............................*Ribes*. 8

1. Genus **SUL-LI-VAN'-TI-A.**—Plants low, reclined-spreading; leaves rounded, cut-toothed, or slightly lobed; flowers small, white, in a branched loosely-cymose panicle on a nearly leafless stem; petals 5, oblanceolate, stamens 5.

Sul-li-van'-ti-a sul-li-van'-ti-i (T. & G.) Britt. (*S. ohionis* T. & G.) **Sullivantia.**—On shaded cliffs; common.

2. Genus **SAX-IF'-RA-GA.** Herbs with clustered root-leaves, those of the stem alternate; calyx free or cohering with the base of the ovary; stamens 10.

Sax-if'-ra-ga vir-gin-i-en'-sis Mx. **Early Saxifrage.**—Low, 4–9 in. high; leaves obovate or oval-spatulate, crenate-toothed; lobes of the calyx not half as long as the white petals.

Sax-if'-ra-ga penn-syl-van'-i-ca L. **Swamp Saxifrage.**—Large, 1–2 ft. high; leaves oblanceolate, 4–8 in. long, obscurely toothed, narrowed at base into a short and broad petiole; lobes of the calyx about the length of the greenish small petals.

3. Genus **TI-A-REL'-LA.** Calyx bell-shaped, petals 5, with claws; stamens 10, long and slender; flowers white in a simple raceme.

Ti-a-rel'-la cor-di-fo'-li-a L. **False Mitre-wort.**—Leaves heart-shaped, sharply lobed and toothed, downy beneath; stem leafless or rarely with 1 or 2 leaves.

4. Genus **HEU'-CHE-RA.**—Calyx bell-shaped, petals 5, spatulate, small; leaves round-cordate, principally from the rootstock; flowers small, greenish or purplish, in a long and narrow panicle.

Heu'-che-ra a-mer-i-ca'-na L. **Alum-root.**—Stems 2 or 3 ft. high, glandular; the spatulate petals not longer than the calyx-lobes.

5. Genus **MI-TEL'-LA.**—Calyx short, petals 5, slender, pinnatifid, stamens 10, included; flowers small in a slender raceme or spike.

Mitella diphylla L. **Mitre-wort: Bishop's Cap.**—Hairy; leaves cordate, acute, somewhat 3 to 5-lobed, those of the stem 2, opposite, nearly sessile; flowers white, numerous.

Mi-tel'-la nu'-da L. **Naked Mitre-wort.**—Small and slender; leaves rounded or reniform, deeply and doubly crenate, stem usually leafless; flowers few, greenish. Northward; reported for Ohio.

Dicotyls or Exogenous Plants.

6. Genus **CHRY-SO-SPLE′-NI-UM.**— Calyx-tube coherent with the ovary, the blunt lobes 4–5, yellow within; petals none, stamens 8–10.

Chry-so-sple′-ni-um a-mer-i-ca′-num Schw. **Golden Saxifrage.**— Small, decumbent; leaves principally opposite, roundish or sub-cordate; flowers inconspicuous, nearly sessile; in springs or streams.

7. Genus **HY-DRAN′-GE-A.**— Calyx-tube hemispherical, 8–10 ribbed, petals ovate; flowers in compound cymes, the marginal flowers usually sterile and radiant, consisting of a dilated colored calyx.

Hy-dran′-ge-a ar-bo-res′-cens L. **Wild Hydrangea.**—Shrub, 1–8 ft. high.

8. Genus **RI′-BES.**—Low, sometimes prickly shrubs with alternate and palmately-lobed leaves; calyx coherent with the ovary, petals perigynous.

A. *Mostly with thorns and prickles.*
  b. *Peduncles 1 to 3-flowered, leaves 3 to 5-lobed.*

Ri′-bes cyn-os′-ba-ti L. **Wild Gooseberry.**—Calyx-lobes shorter than the tube; berries mostly prickly; stamens not longer than the bell-shaped calyx.

Ri′-bes ro-tun-di-fo′-li-um Mx. **Wild Gooseberry.**— Calyx-lobes longer than the short tube; peduncles short; filaments long, somewhat exceeding the narrowly oblong-spatulate calyx-lobes; fruit smooth.

Ri′-bes ox-y-can-thoi′-des L. **Wild Gooseberry.**— Calyx-lobes as in the last; peduncles very short; stamens usually scarcely exceeding the rather broadly-oblong calyx-lobes; fruit smooth.

  b. *Flowers several in a nodding raceme, leaves 3 to 5-parted.*

Ri′-bes la-cus-tre (Pers.) Poir. **Wild Gooseberry.** - Young stems with bristly prickles and weak thorns; fruit bristly, small.

A. *Without thorns or prickles.*

Ri′-bes flor′-i-dum L'Her. **Wild Black Currant.** — Racemes drooping, downy; fruit black; leaves lobed, doubly serrate.

Ri′-bes ru′-brum L. **Red Currant.**— Racemes from lateral buds distinct from the leaf-buds, drooping, fruit red; leaves lobed and serrate.

XXXIV. Order **HAM-A-ME-LI-DA′-CE-Æ. WITCH-HAZEL FAMILY.** — Shrubs or trees; leaves alternate, simple; flowers clustered in spikes or heads; the calyx cohering with the base of the ovary.

Petals 4, strap-shaped, perfect stamens and scales each 4........*Hamamelis*. 1
Petals none, calyx rudimentary, stamens numerous...........*Liquidamber*. 2

1. Genus **HAM-A-ME'-LIS.** — Flowers in little axillary clusters or heads, yellow, petals long and narrow; tall shrubs.

Ham-a-me'-lis vir-gin-i-a'-na L. Leaves obovate or oval; blossoming late in autumn, the seeds maturing the next summer.

2. Genus **LIQ-UID-AM'-BER.** Flowers usually monœcious, in globular clusters; fruit a spherical hard catkin or head.

Liq-uid-am'-ber sty-ra-cif'-lu-a. Sweet Gum; Bilsted. — Leaves deeply 5 to 7-lobed, smooth and shining. A large tree of extreme southern Ohio.

XXXV. Order **PLAT-A-NA'-CE-Æ. PLANE-TREE FAMILY.** Trees; leaves large, palmately lobed, alternate; flowers in spherical heads.

1. Genus **PLAT'-A-NUS.** — Flowers monœcious in naked spherical heads, destitute of calyx and corolla; leaves palmately lobed.

Plat'-a-nus oc-ci-den-ta'-lis L. **Sycamore.** Our largest tree; on alluvial banks; the bark deciduous in brittle plates.

XXXVI. Order **RO-SA'-CE-Æ. ROSE FAMILY.** — Flowers with 5 petals (rarely wanting) and numerous distinct stamens inserted on the calyx; leaves alternate, stipulate.

<pre>
Ovary superior and not enclosed in the calyx-tube at maturity a .
Ovaries inferior or enclosed on the calyx-tube d .
a. Herbs c .
a. Shrubs or trees b .
   b. Shrub; leaves palmately lobed, pods inflated..................*Opulaster*. 1
   b. Shrub; leaves simple, pistils about 5, pods not inflated..........*Spiræa*. 2
   b. Shrub; leaflets 5-7, crowded, petals yellow i .
   b. Biennial soft woody stems; lvs. simple or compound, pistils numerous h).
   b. Trees or shrubs; leaves simple, pistil solitary, fruit a drupe k .
c. Leaves thrice pinnate, flowers in a long compound panicle............*Aruncus*. 3
c. Leaves, etc., not as above i .
   d. Leaves pinnate f .
   d. Leaves simple e .
e. Cymes simple and umbel-like, fruit a fleshy pome, globular, sunk in
   at the attachment of the stalk, carpels 2-5, papery or cartilaginous,
   2-seeded..*Pyrus*. 4
e. Not as above g .
   f. Pistils becoming achenes, enclosed in the fleshy calyx-tube k .
</pre>

Dicotyls or Exogenous Plants. 61

```
f. Carpels 2-5, forming a berry-like pome.................................Sorbus.  5
g. Flowers cymose, fruit berry-like, small....................................Aronia.  6
g. Flowers racemose, pome berry-like, 10-celled.........................Amelanchier.  7
g. Flowers corymbose, pome drupe-like, 1-5 bony seeds................Cratægus.  8
    h. Pistils numerous, fleshy in fruit......................................Rubus.  9
i. Leaflets 3, obovate-cuneate, coarsely serrate, leaves radical, flowers
    white ...............................................................Fragaria. 10
i. Leaflets 3-21, leaves radical and cauline; flowers yellow or white, calyx
    5-cleft, with 5 bractlets at the sinuses, styles deciduous.........Potentilla. 11
i. Leaflets 5-7, petals purple, short; calyx dark-purple inside with 5
    bractlets..........................................................Comarum. 12
i. Leaflets 3, broadly cuneate, cut-toothed, leaves radical; flowers yel-
    low ............................................................Waldsteinia. 13
i. Root-leaves simple and rounded with minute leaflets on the petiole
    below, or 3 to 5-lobed, or pinnate, 3-5 leaflets; styles long and per-
    sistent; calyx with 5 bractlets......................................Geum. 14
i. Leaves interruptedly pinnate, terminal leaflet large, 7 to 9-parted,
    calyx 5-cleft, no bractlet..........................................Ulmaria. 15
k. Leaves compound, calyx-tube urn-shaped, fleshy in fruit............Rosa. 16
k. Leaves simple, calyx deciduous after flowering....................Prunus. 17
```

1. Genus **OP-U-LAS'-TER.**— Carpels 1-5, inflated; leaves roundish, somewhat 3-lobed; flowers white in umbel-like corymbs.

Op-u-las'-ter op-u-li-fo'-li-us (L.) Kuntze. (*Physiocarpus opulifolius* Maxim.) **Nine Bark.** Shrub 4-10 ft. high, with long recurved branches; pods purplish, conspicuous.

2. Genus **SPI-RÆ'-A.**—Petals 5, obovate; follicles (pods) 5-8, not inflated; flowers white or rose-color, in corymbs or panicles; leaves simple; shrubs.

Spi-ræ'-a cor ym-bo'-sa Raf. (*S. betulæfolia* var. *corymbosa* Wats.) **Meadow Sweet.** Nearly smooth, 1-2 ft. high; leaves oval or ovate, cut-toothed toward the apex; corymbs large; flowers white.

Spi-ræ'-a sal-i-ci-fo'-li-a L. **Common Meadow Sweet.**—Nearly smooth, 2-3 ft. high; leaves cuneate-lanceolate, simply or doubly serrate; flowers in a crowded panicle, white or flesh color.

Spi-ræ'-a to-men-to'-sa L. **Hardhack: Steeple-Bush.**—Stems and lower surface of the ovate or oblong serrate leaves *very* woolly; flowers in short racemes crowded in a dense panicle, rose-color, or occasionally white.

3. Genus **A-RUN'-CUS.**—Flowers diœcious, whitish, in many slender spikes, disposed in a long compound panicle; leaves 3-pinnate; pedicels reflexed in fruit.

A-run'-cus a-run'-cus (L.) Karst. (*Spiræa aruncus* L.) **Goat's Beard.**—A smooth, tall, perennial herb.

4. Genus **PY'-RUS.**—Trees or shrubs, with flowers in corymbed cymes, large, fragrant, rose color in the native species; pome globular. Besides the species named below the genus also includes the cultivated Apple (*Pyrus malus* L.) and the Pear (*Pyrus communis* L.)

Py'-rus cor-o-na'-ri-a L. **Crab-Apple.**—Leaves ovate, often sub-cordate, cut-serrate or lobed, soon glabrous; styles united at base.

Py'-rus an-gus-ti-fo'-li-a Ait. **Crab-Apple.**—Resembling the last, but the leaves oblong or lanceolate, often acute at base, mostly toothed, glabrous; styles distinct; blooming earlier. apr. 15

5. Genus **SOR'-BUS.**—Trees or tall shrubs with flowers in compound cymes; styles separate, pome berry-like, small; leaves pinnate, leaflets rather numerous.

Sor'-bus sam-bu-ci-fo'-li-a (C. & S.) Roem. (*Pyrus sambucifolia* Ch. & Schl.) **Mountain Ash.**—Leaflets oblong, oval or lanceolate; berries 4 lines broad.

6. Genus **A-RO'-NI-A.**—Shrub, leaves simple, the midrib glandular along the upper side; cymes compound, fruit berry-like, small.

A-ro'-ni-a ar-bu-ti-fo'-li-a (L.) Ell. (*Pyrus arbutifolia* L. f.) **Chokeberry.**—Leaves oblong or oblanceolate, finely glandular-serrate, tomentose beneath; cyme tomentose; fruit small, red or purple. Swamps and damp thickets.

A-ro'-ni-a ni'-gra (Willd.) Britt. (*Pyrus arbutifolia* var. *melanocarpa* Hk.) **Chokeberry.**—Differs from the preceding in being nearly smooth throughout; with larger black fruit.

7. Genus **AM-E-LAN'-CHI-ER.**—Calyx-lobes downy within, petals oblong, elongated; the pome berry-like, 10-celled; small trees with sharply serrate leaves and white racemose flowers.

Am-e-lan'-chi-er can-a-den'-sis (L.) Medic. **Shad-bush; Service-berry; June-berry.**—Tree 10–30 ft. high; leaves ovate to ovate-oblong, mostly sub-cordate at base, 1–3½ in. long; flowers in drooping racemes, petals 6–8 lines long.

Am-e-lan'-chi-er bo-try-a'-pi-um (L.f.) DC. (*A. canadensis* var. *oblongifolia* T. & G.) **Shad-bush; Service-berry: June-berry.** —Smaller tree or shrub, 6-10 ft. high; leaves oblong or sometimes rather broadly-elliptical, mostly rounded at base, 1-2 in. long; flowers in denser and shorter racemes; petals 3-4 lines long.

8. Genus **CRA-TÆ'-GUS.** — Thorny shrubs or small trees, with simple and mostly lobed leaves; petals roundish; pome drupe-like, containing 1-5 bony seeds.

A. *Leaves broad.*

Cra-tæ'-gus coc-cin'-e-a L. **Scarlet Haw; White Thorn.**— Branches reddish, spines stout, villous-pubescent on the shoots, glandular peduncles and calyx; leaves on slender petioles, round-ovate, cuneate or sub-cordate at base, glandular-toothed, sometimes cut-lobed; *flowers ½ in. broad;* fruit coral-red, globose or obovate, ½ in. broad.

Cra-tæ'-gus ma-cra-can'-tha Lodd. (*C. coccinea* var. *macracantha* Dudley.) **Large-spine Hawthorn.** Differs from the preceding in having longer, bright chestnut-brown spines, thicker leaves, sub-coriaceous, cuneate at base, *on stout petioles*, often deeply incised; cymes broader, flowers and fruit rather larger.

Cra-tæ'-gus mol'-lis (T. & G.) Scheele. (*C. coccinea* var. *mollis* T. & G.) **Black Thorn; Scarlet Haw.**— Shoots densely pubescent, leaves large, broadly ovate, more or less pubescent beneath; *flowers 1 in. broad or more;* fruit bright scarlet with a light bloom, 1 in. broad; blooming very early.

Cra-tæ'-gus to-men-to'-sa L. **Black Thorn.**— Branches gray; shoots, peduncles and calyx villous-pubescent; leaves large, pale, prominently veined, densely pubescent beneath, *contracted into a margined petiole; flowers small, ill-scented;* fruit dull red, ½ in. broad; blooming late.

A. *Leaves narrower.*

Cra-tæ'-gus punc-ta'-ta Jacq. **Black Thorn.** — Leaves mostly wedge-obovate, not shining, attenuate and entire near the base, unequally toothed toward the apex, rarely lobed; fruit globose, 1 in. broad.

Cra-tæ'-gus crus-gal'-li L. **Cockspur Thorn.**— Thorns slender, often 4 in. long; *leaves thick, coriaceous, dark-green, shining above*, wedge-obovate and lanceolate, serrate above the middle.

64   *Spring Flora of Ohio.*

9. Genus **RU'-BUS.**—Calyx 5-parted, without bractlets; achenes usually many on a spongy or succulent receptacle; perennial herbs or somewhat shrubby.

A. *Fruit falling off whole from the dry receptacle when ripe, or consisting of few grains which fall separately.*

b. *Leaves simple, flowers large, prickles none.*

Ru'-bus o-do-ra'-tus L. **Purple-Flowering Raspberry.**—Leaves 3-5-lobed; flowers showy, 2 in. broad, purple rose-color.

b. *Leaflets 3-5, petals small, erect, white.*

Ru'-bus a-mer-i-ca'-nus (Pers.) Brit. (*R. triflorus* Rich.) **Dwarf Raspberry.**—Stems ascending (6-12 in. high) or trailing, not prickly; leaflets 3 (or 5), acute at both ends, coarsely doubly-serrate, thin; peduncle 1-3-flowered.

Ru'-bus stri-go'-sus Mx. **Red Raspberry.**—Stems upright, beset with stiff straight bristles (or a few weak hooked prickles), glandular when young, somewhat glaucous; petals as long as the sepals; fruit light-red.

Ru'-bus oc-ci-den-ta'-lis L. **Black Raspberry.**—Glaucous all over; stems recurved, with hooked prickles, not bristly; petals shorter than the sepals; fruit purple-black.

A. *Fruit not separating from the juicy, prolonged receptacle.*

c. *Stems upright or reclining, branchlets and lower surface of leaves hairy and glandular.*

Ru'-bus vil-lo'-sus Ait. **Common or High Blackberry.** Stems furrowed, 1-6 ft. high, armed with stout, curved prickles; flowers racemed, numerous, petals obovate-oblong.

Ru'-bus vil-lo'-sus fron-do'-sus Torr. **Blackberry.** Differs from the last in being smoother, much less glandular, flowers more corymbose, bracts leafy, petals roundish.

c. *Stems trailing or procumbent; leaflets nearly or quite smooth.*

Ru'-bus can-a-den'-sis L. **Dewberry: Low Blackberry.**—Shrubby, extensively trailing, slightly prickly, leaflets 3 (or 5 or 7), oval, or ovate lanceolate.

Ru'-bus his'-pi-dus L. **Running Swamp-Blackberry.**—Stem slender, scarcely woody, extensively procumbent, beset with small reflexed prickles; leaflets obovate, obtuse; fruit of a few grains.

c. *Stems upright, branchlets and lower side of leaves whitish-woolly.*

Dicotyls or Exogenous Plants. 65

Ru'-bus cu-ne-i-fo'li-us Ph. **Sand Blackberry.**—Shrubby, 1–3 ft. high, armed with stout, recurved prickles; peduncles 2–4-flowered; petals large.

10. Genus **FRA-GA'-RI-A.**—Receptacle in fruit much enlarged and conical, becoming pulpy and scarlet; leaves radical; leaflets 3, coarsely serrate.

A. *Achenes imbedded in the deeply-pitted fruiting receptacle.*

Fra-ga'-ri-a vir-gin-i-a'-na Mill. **Wild Strawberry.**—Receptacle usually with a narrow neck, calyx becoming erect after flowering, and connivent over the hairy receptacle when sterile; leaflets firm; hairs of the scape and especially of the pedicels *silky and appressed.*

Fra-ga'-ri-a vir-gin-i-a'-na il-li-no-en'-sis Prince. **Wild Strawberry.**—A coarser or larger plant, *the villous hairs* of the scape and pedicels widely spreading.

A. *Achenes not sunk in pits, superficial on the glabrous receptacle.*

Fra-ga'-ri-a ves'-ca L. **Strawberry.**—Calyx remaining spreading or reflexed; hairs on the scape mostly widely spreading, on the pedicels appressed; leaflets thin, even the upper face strongly marked by the veins.

Fra-ga'-ri-a a-mer-i-ca'-na (Porter) Britt. **Strawberry.**—Much like the preceding, but plant softly villous, scape and peduncles slender, flowers small; fruit of a light pink color.

11. Genus **PO-TEN-TIL'-LA.** — Herbs or rarely shrubs; leaves compound, flowers solitary or cymose; calyx 5-cleft, with 5 bractlets, appearing 10-cleft; receptacle dry.

A. *Petals whitish or cream-color, broad, surpassing the calyx.*

Po-ten-til'-la ar-gu'-ta Ph. **Cinque-foil; Five-finger.**—Stems erect, 1–4 ft. high, leaflets 7–11, downy beneath.

A. *Petals pale-yellow, not surpassing the calyx.*

Po-ten-til'-la mon-spe-li-en'-sis L. (*P. norvegica* L.) **Cinque-foil; Five-finger.**—Stout, erect, hirsute, ½–2 ft. high; leaves ternate, calyx large.

A. *Petals bright or light-yellow, larger than the lobes of the calyx.*

b. *Inflorescence cymose.*

Po-ten-til'-la ar-gen'-te-a L. **Silvery Cinque-foil.**—Leaves palmate, leaflets 3–5; white woolly; herb.

Po-ten-til'-la fru-ti-co'-sa L. **Shrubby Cinque-foil.**—Stem erect, shrubby, 1–4 ft. high; leaves pinnate, leaflets 5–7, oblong-lanceolate, entire.

Po-ten-til'-la rec'-ta Willd. **European Cinque-foil; Five-finger.**—Coarse, erect, simple, hairy, 1–3 ft. high; leaflets 5–7, narrowly wedge-oblong, coarsely toothed; stipules large, cleft.

  b. *Peduncles axillary, solitary, 1-flowered.*

Po-ten-til'-la an-se-ri'-na L. **Silver-weed.**— Spreading by slender, many-jointed runners, white tomentose and silky villous; leaves all radical, pinnate.

Po-ten-tli-la can-a-den-sis L. **Common Cinque-foil** or **Five-finger.**— Stems slender and decumbent or prostrate, or sometimes erect; leaves ternate, but the lateral leaflet parted, appearing quinate.

12. Genus **CO'-MA-RUM.** — Petals 5, much smaller than the sepals; calyx deeply 5-cleft, bractlets 5; receptacle ovate, spongy.

Co'-ma-rum pa-lus'-tre L. (*Potentilla palustris* Scop.) **Marsh Five-finger.**—Petals purple, calyx 1 in. broad, purple inside; stems ascending from a decumbent base; leaflets 5–7.

13. Genus **WALD-STEI'-NI-A.** — Calyx-tube inversely conical with 5 often minute and deciduous bractlets; achenes 2–5; leaves chiefly radical, leaflets 3, broadly cuneate, cut-toothed.

Wald-stei'-ni-a fra-ga-ri-oi'-des (Mx.) Tratt. **Barren Strawberry.**—Flowers small, yellow, on bracted scapes.

14. Genus **GE'-UM.**—Calyx deeply 5-cleft, usually with 5 small bractlets at the sinuses; achenes numerous, the long persistent styles either straight or jointed; leaves pinnate or lyrate.

Ge'-um can-a-den'-se Jacq. (*G. album* Gmel.) **Avens.**— Petals white or pale greenish-yellow; styles jointed and bent near the middle, the upper part deciduous and mostly hairy; head of fruit sessile in the calyx.

Ge'-um ver'-num (Raf.) T. & G. **Avens.**—Petals yellow, about the length of the calyx; styles smooth; head of fruit conspicuously stalked in the calyx.

Ge'-um ri-va'-le L. **Water** or **Purple Avens.**—Petals dilated-obovate; purplish-orange; styles jointed and bent in the middle, the upper joint plumose; head of fruit stalked in the brown purple calyx.

15. Genus **UL-MA'-RI-A.**—Perennial herbs with pinnate leaves and panicled cymose flowers; stipules kidney-form; calyx 5-cleft, short; pods 5-8.

Ul-ma'-ri-a ru'-bra Hill. (*Spiræa lobata* Jacq.) **Queen of the Prairie.**—Plant 2-8 ft. high; flowers deep peach-blossom color, the petals and sepals often four.

16. Genus **RO'-SA.**—Calyx-tube urn-shaped, contracted at the mouth, becoming fleshy in fruit; ovaries hairy, becoming bony in fruit; shrubs with odd-pinnate leaves.

A. *Styles cohering in a protruding column.*

Ro'-sa se-tig'-e-ra Mx. **Climbing or Prairie Rose.**—Stems climbing, armed with nearly straight scattered prickles; leaflets 3-5, ovate.

A. *Styles distinct, sepals connivent after flowering, persistent.*

Ro'-sa blan'-da Ait. **Wild Rose.**—Stems 1-3 ft. high, wholly unarmed (or rarely with prickles); leaflets 5-7, cuneate at base, serrate.

A. *Styles distinct, sepals spreading after flowering and deciduous.*

b. *Leaflets mostly finely many-toothed.*

Ro'-sa car-o-li'-na L. **Carolina Rose.**—Stems usually tall, 1-7 ft. high; spines stout, straight or usually more or less curved; leaflets 5-9, usually narrowly oblong and acute at each end, or broader.

b. *Leaflets coarsely toothed.*

Ro'-sa lu'-ci-da Ehrh. **Wild Rose.**—Stems often tall and stout (a few inches to 6 ft.); spines stout and usually more or less hooked; stipules more or less dilated; leaflets mostly 7; outer sepals frequently with 1 or 2 lobes.

Ro'-sa hu'-mi-lis Marsh. **Wild Rose.**—Stems usually low, 1-3 ft. high and more slender, less leafy; spines straight, slender, spreading or sometimes reflexed; stipules narrow, rarely somewhat dilated; outer sepals always more or less lobed.

b. *Leaflets doubly serrate.*

Ro'-sa ru-bi-gin-o'-sa L. **Sweet Brier.**—Stems with stout recurved spines; leaflets *densely resinous beneath and aromatic;* the short pedicels and pinnatifid sepals hispid.

17. Genus **PRU'-NUS.**—Calyx 5-cleft, petals 5, spreading; stamens 15-20, pistil solitary; drupe fleshy with a bony stone; trees or shrubs. Besides the species named below, this genus also includes the cultivated Plums, Cherries, Peach, Apricot, etc.

A. *Flowers with or preceding the leaves in umbel-like clusters.*

Pru'-nus a-mer-i-ca'-na Marsh. **Wild Yellow** or **Red Plum.** — Tree thorny, 8-20 ft. high; leaves conspicuously pointed, coarsely and doubly serrate, very veiny; fruit ½-¾ in. in diameter.

Pru'-nus penn-syl-van'-i-ca L. f. **Wild Red Cherry.**— Tree 20-30 ft. high; leaves pointed, finely and sharply serrate, shining green and smooth both sides; fruit light red, very small, sour.

A. *Flowers in racemes, appearing after the leaves.*

Pru'-nus vir-gin-i-a'-na L. **Choke Cherry.**—A tall shrub or occasionally a tree; leaves oval, oblong or obovate, abruptly pointed, very sharply, often doubly serrate with slender teeth, thin; petals roundish.

Pru'-nus se-rot'-i-na Ehrh. **Wild Black Cherry.**—A large tree; leaves oblong or lanceolate-oblong, taper-pointed, serrate with incurved short and callous teeth, thickish, shining above; petals obovate.

XXXVII. Order **LE-GU-MIN-O'-SÆ.** **PEA FAMILY.**—Flowers mostly papilionaceous, or sometimes regular; stamens 10 (rarely 5 or many), distinct or united by their filaments; pistil simple, free; fruit a legume.

```
Trees or shrubs  a.
Herbs  c.
a. Leaves compound  b.
a. Leaves simple, round-cordate; flowers preceding the leaves............Cercis. 1
   b. Trees thorny, leaves 1-2-pinnate, leaflets small; calyx short, 3
      to 5-cleft; flowers not papilionaceous.................Gleditschia. 2
   b. Trees not thorny; leaves all doubly pinnate, very large; calyx
      elongated-tubular below, 5-cleft; flowers not papilionace-
      ous ..........................................................Gymnocladus. 3
   b. Trees and shrubs, leaves pinnate; flowers papilionaceous f.
c. Stamens 10 distinct, pod inflated............................Baptisia. 4
c. Stamens monadelphous or diadelphous  9 & 1) (d.
   d. Leaves palmate (digitate), leaflets 7-11.....................Lupinus. 5
   d. Leaves trifoliate  e'.
   d. Leaves abruptly pinnate  g'.
e. Flowers racemed or spiked, pods curved or coiled..............Medicago. 6
e. Flowers racemed, pods coriaceous, wrinkled.....................Melilotus. 7
e. Flowers in heads, pods membranaceous..........................Trifolium. 8
   f. Woody twiner, pod thickish.................................Kraunhia. 9
   f. Trees or shrubs, pod flat..................................Robinia. 10
g. Leaves terminated by a bristle or tendril.....................Vicia. 11
```

Dicotyls or Exogenous Plants.

1. Genus **CER'-CIS.**—Flowers red-purple, preceding the leaves; leaves round-cordate, pointed.

Cer'-cis can-a-den'-sis L. **Red-bud.**—Small trees in rich soil.

2. Genus **GLE-DITSCH'-I-A.**—Thorny trees, with abruptly 1-2-pinnate leaves; flowers inconspicuous, in small spikes.

Gle-ditsch'-i-a tri-a-can'-thos L. **Honey Locust.**—Pods linear-elongated, 1-1½ ft. long.

3. Genus **GYM-NOC'-LA-DUS.**—Flowers diœcious or polygamous, regular; petals 5, inserted on the summit of the calyx-tube; a large tree with stout branchlets and large, uneqally 2-pinnate leaves.

Gym-noc'-la-dus di-oi'-cus (L.) Koch. (*G. canadensis* Lam.) **Kentucky Coffee-tree.**—Leaves 2-3 ft. long; pod 2-10 in. long, 2 in. broad.

4. Genus **BAP-TIS'-I-A.**—Calyx 4-5-toothed; stamens 10, distinct; pods stalked in the persistent calyx, roundish or oblong; leaflets 3.

Bap-tis'-i-a tinc-to'-ri-a R. Br. **Wild Indigo.**—Smooth and slender, rather glaucous; racemes many, short and loose; flowers yellow; stipules minute.

Bap-tis'-i-a leu-can'-tha T. & G. **White False Indigo.** — Smooth, tall and stout; raceme elongated, 1-2 ft. long, erect, flowers white; stipules lanceolate, as long as the petioles.

Bap-tis'-i-a aus-tra'-lis R. Br. **Blue False Indigo.**—Resembling the last, but flowers indigo-blue and the stalk of the pods only about the length of the calyx; flowering later.

5. Genus **LU-PI'-NUS.**—Calyx deeply 2-lipped; stamens monadelphous; leaves palmate; flowers showy, in a long, terminal raceme.

Lu-pi'-nus per-en'-nis L. **Wild Lupine.** Flowers purplish-blue, rarely pale; stem erect, 1 ft. high; leaves soft-downy.

6. Genus **MED-I-CA'-GO.**—Flowers nearly as in Trifolium (Clover), but in racemes or spikes; pods scythe-shaped, incurved or coiled.

Med-i-ca'-go sa-ti'-va L. **Lucerne; Alfalfa.**—Upright, smooth; flowers purple, racemed; pods spirally twisted.

Med-i-ca'-go lu-pu-li'-na L. **Black Medick; Nonesuch.** — Procumbent, pubescent; flowers yellow, in *short spikes*; pods kidney-form.

7. Genus **MEL-I-LO'-TUS.** Flowers nearly as in Trifolium (Clover), but small and in spike-like racemes; pod ovoid, coriaceous, wrinkled.

Mel-i-lo'-tus of-fic-i-na'-lis Willd. **Yellow Melilot: Sweet Clover.** — Upright, 2-4 ft. high; leaflets obtuse; corolla yellow.

Mel-i-lo'-tus al'-ba Lam. **White Melilot: Sweet Clover.** — Leaflets truncate; corolla white.

8. Genus **TRI-FO'-LI-UM.** — Calyx 5-cleft, the teeth bristle-form; the claws of the petals more or less united below with the stamen-tube.

A. *Flowers sessile, not reflexed, in dense heads or spikes, purple, purplish, roseate or whitish.*

Tri-fo'-li-um ar-ven'-se L. **Rabbit-foot or Stone Clover.** — Annual, silky, 5-10 in. high; calyx-teeth silky-plumose, longer than the whitish corolla; leaflets oblanceolate. Naturalized from Europe.

Tri-fo'-li-um pra-ten'-se L. **Red Clover.** — Stems ascending, somewhat hairy; leaflets oval or obovate, often notched; calyx shorter than the tubular corolla.

Tri-fo'-li-um in-car-na'-tum L. **Crimson Clover.** — Leaflets ovate, orbicular, obtuse or obcordate, crenate, villous; spike oblong, pedunculate. Cultivated and often escaped.

A. *Flowers pedicelled, in umbel-like round heads on a naked peduncle, their short pedicels reflexed when old; corolla white or rose-color, persistent.*

Tri-fo'-li-um sto-len-if'-er-um Muhl. **Running Buffalo Clover.** — Leaflets broadly obovate or obcordate; flowers white, tinged with purple; pods 2-seeded.

Tri-fo'-li-um re'-pens L. **White Clover.** — Leaflets inversely heart-shaped or merely notched; corolla white; pods about 4-seeded.

Tri-fo'-li-um hy'-bri-dum L. **Alsike Clover.** — Resembling White Clover (*T. repens*), but the stems erect or ascending, not rooting at the nodes; flowers rose-tinted.

A. *Flowers short-pedicelled in close heads, reflexed when old; corolla yellow; annuals, not flowering early.*

Dicotyls or Exogenous Plants. 71

Tri-fo'-li-um pro-cum'-bens L. **Low Hop Clover.** — Stems spreading or ascending; leaves pinnately 3-foliate, leaflets wedge-obovate, notched at the end; stipules ovate.

9. Genus **KRAUN'-HI-A.** (*Wistaria.*) — Flowers large, showy, lilac-purple, in dense racemes; pods elongated; leaflets 9-13; woody twiners.

Kraun'-hi-a fru-tes'-cens (L.) Greene. (*Wistaria frutescens* Poir.) — Cultivated; indigenous southward.

10. Genus **RO-BIN'-I-A.** — Calyx slightly 2-lipped; trees or shrubs, often with prickly spines for stipules; flowers showy in hanging axillary racemes. Cultivated and sometimes escaped.

Ro-bin'-i-a pseu-da-ca'-ci-a L. **Common Locust.** — Branchlets naked; racemes slender, loose, flowers white, fragrant; pod smooth.

Ro-bin'-i-a vis-co'-sa Vent. **Clammy Locust.** — Branchlets and leaf-stalks clammy; flowers crowded in oblong racemes, tinged with rose-color.

Ro-bin'-i-a his'-pi-da L. **Bristly Locust.** — Shrub 3-8 ft. high; branchlets and stalks bristly, or sometimes nearly naked; flowers large and deep rose-color.

11. Genus **VI'-CI-A.** — Herbs, mostly climbing by the tendril at the end of the pinnate leaves; stipules half-sagittate; wing of the corolla adhering to the middle of the keel; not flowering early.

A. *Annual; flowers 1 or 2 in the axils, or elongated peduncles 3-6 flowered.*

Vi'-ci-a sa-ti'-va L. **Common Vetch** or **Tare.** — Flowers 1 or 2 in the axils, nearly sessile, large, violet-purple; leaflets 5-7 pairs, notched or mucronate at apex. Introduced from Europe.

Vi'-ci-a hir-su'-ta Koch. **Hairy Vetch.** — Peduncles 3-6-flowered; leaflets 6-8 pairs, truncate. Naturalized from Europe.

A. *Perennial, indigenous species; peduncles 4-many-flowered.*

Vi'-ci-a crac'-ca L. **Vetch.** — Downy-pubescent; leaflets 20-24, oblong-lanceolate, strongly mucronate; spikes densely many-flowered, 1-sided; flowers blue, turning purple.

Vi'-ci-a car-o-li-ni-a'-na Walt. **Vetch.** — Nearly smooth; leaflets 8-24, oblong, obtuse, scarcely mucronate; peduncles loosely-flowered; flowers whitish, the keel tipped with blue.

Vi'-ci-a a-mer-i-ca'-na Muhl. **Vetch.**—Glabrous; leaflets 10-14, elliptical or ovate-oblong, very obtuse; peduncles 4-8-flowered; flowers purplish, 8 lines long.

XXXVIII. Order **GE-RA-NI-A'-CE-Æ. GERANIUM FAMILY.**—Herbs; flowers regular, 5-merous; stamens somewhat united; ovary deeply-lobed; carpels 5, separating from the styles when mature; leaves lobed or divided.

Stamens 10 rarely 5, base of styles recurving, naked inside......*Geranium.* 1
Stamens with anthers 5, styles in fruit twisting spirally, bearded. *Erodium.* 2

1. Genus **GE-RA'-NI-UM.**—Stamens all with perfect anthers, the 5 longer with glands at their base (sometimes only 5 stamens).

A. *Flowers large; rootstock perennial.*

Ge-ra'-ni-um mac-u-la'-tum L. **Wild Cranesbill.**—Leaves about 5-parted, the divisions wedge-shaped, lobed and cut at the end; petals entire, light-purple, bearded on the claw.

A. *Flowers small; root biennial or annual.*

b. *Leaves ternately much-dissected; heavy-scented.*

Ge-ra'-ni-um ro-ber-ti-a'-num L. **Herb Robert.**—Diffuse, strong-scented; sepals awned, shorter than the red-purple petals.

b. *Leaves palmately-lobed or dissected.*

Ge-ra'-ni-um car-o-li-ni-a'-num L. **Cranesbill.**—Stems at first erect, diffusely branched from the base; leaves about 5-parted, the divisions cleft and cut into oblong-linear lobes; sepals awn-pointed; variable.

The following European species are occasionally found: *G. dissectum* more slender and spreading, with narrower lobes to the crowded leaves, seeds finely and deeply pitted ; *G. rotundifolium* with the habit of the next but seed of the last, villous with long white hairs ; *G. pusillum* stems procumbent, slender, the leaves round-kidney form, sepals awnless, seeds smooth ; *G. molle* like the last) more pubescent, seed slightly striate, and *G. columbinum* with peduncles and pedicels filiform and elongated, seeds nearly as in G. dissectum .

2. Genus **E-RO'-DI-UM.**—The 5 shorter stamens sterile or wanting; styles in fruit twisting spirally, bearded inside.

E-ro'-di-um ci-cu-ta'-ri-um L'Her. **Storks-bill.**—Stems low, spreading; leaves pinnate, leaflets sessile, 1-2-pinnatifid.

XXXIX. Order **OX-AL-I-DA'-CE-Æ. WOOD-SORREL FAMILY.**—Herbs; leaves trifoliate; flowers 5-merous; fruit a 5-celled pod.

Dicotyls or Exogenous Plants.

1. Genus **OX'-A-LIS.**—Sepals 5, persistent; petals 5, sometimes united at base; stamens 10, monadelphous at base; styles 5, distinct; leaves radical or alternate; juice sour.

A. *Stemless, leaves and scapes from a bulb.*

Ox'-a-lis vi-o-la'-ce-a L. **Violet Wood-sorrel.**—Scapes umbellately several-flowered, 5–9 in. high; petals violet.

A. *Stems leafy, branching, peduncles axillary, flowers yellow.*

Ox'-a-lis cor-nic-u-la'-ta L. **Wood-sorrel.** — Erect or procumbent; strigose-pubescent; stipules round or truncate, ciliate.

Ox'-a-lis stric'-ta L. **Wood-sorrel.**—Stem erect, like the preceding, but stipules none; variable.

Ox'-a-lis gran'-dis Small. (*O. recurva* Gray's Man.) **Large Wood-sorrel.**—Differs from the preceding in having leaflets larger; ½–1½ in. broad, usually with a brownish margin; flowers larger, 6–8 lines long.

XL. Order **RU-TA'-CE-Æ. RUE FAMILY.**—Shrubs; leaves trifoliate or pinnate; flowers polygamous or diœcious; pistils 2–5, separate or united.

Flowers diœcious, pistils 2–5, separate, pods fleshy............*Xanthoxylum*. 1
Flowers polygamous, ovary 2-celled, fruit winged all around.........*Ptelea*. 2

1. Genus **XAN-THOX'-Y-LUM.**—Shrubs or trees with pinnate leaves, the stems and often leafstalks prickly; flowers small, greenish or whitish.

Xan-thox'-y-lum a-mer-i-ca'-num Mill. **Prickly Ash; Toothache Tree.** — Flowers appearing before the leaves; bark, leaves and pods very pungent and aromatic.

2. Genus **PTEL'-E-A.**— Shrubs with trifoliate leaves, and greenish-white small flowers in compound terminal cymes.

Ptel'-e-a tri-fo-li-a'-ta L. **Hop Tree: Shrubby Trefoil.** — Leaflets ovate, pointed, downy when young; fruit a samara, nearly orbicular.

XLI. Order **SI-MA-RU-BA'-CE-Æ. AILANTHUS FAMILY.** — Trees; leaves pinnate; sepals 5, flowers panicled; fruit a samara.

1. Genus **AI-LAN'-THUS.** — Sepals 5, more or less united at base, petals 5; fruit a 1-celled, 1-seeded samara; leaves pinnate, 2–4 ft. long.

74   *Spring Flora of Ohio.*

Ai-lan'-thus glan-du-lo'-sa Desf.   **Chinese Tree of Heaven.**—Cultivated and occasionally escaped.

XLII.   Order **PO-LYG-A-LA'-CE-Æ.   MILKWORT FAMILY.** — Flowers irregular, hypogynous, stamens 4–8, diadelphous or monadelphous.

1. Genus **PO-LYG'-A-LA.** — Flowers very irregular; 3 of the sepals small, petals 3, stamens 6 or 8; bitter, low plants. A large genus, most of our species blooming late.

Po-lyg'-a-la pau-ci-fo'-li-a Willd.   **Fringed Polygala.** — Flowers showy, rose-purple (rarely white), conspicuously crested; flowering stems 3–4 in. high.

Po-lyg'-a-la sen'-e-ga L., **Senega Snake-root.** — Flowers white, in a solitary close spike; several stems from thick and hard knotty rootstocks.

XLIII.   Order **EU-PHOR-BI-A'-CE-Æ.   SPURGE FAMILY.**—Herbs with milky juice; staminate and (single) pistillate flowers enclosed in a calyx-like involucre; ovary 3-celled.

1. Genus **EU-PHOR'-BI-A.**—Flowers monœcious, included in a cup-shaped, 4 to 5-lobed involucre resembling a calyx or corolla and usually bearing large, thick glands (with or without petal-like margins) at its sinuses; fertile flower solitary in the middle of the involucre consisting of a 3-lobed and 3-celled ovary.

Eu-phor'-bi-a ob-tu-sa'-ta Ph.   **Spurge.** — Erect, 1–2 ft. high; leaves oblong and spatulate, upper ones cordate at base, floral ones ovate.

Eu-phor'-bi-a cy-par-is'-si-as L.   **Garden Spurge.** — Stems densely clustered, 6–10 in. high; stem-leaves linear, crowded, the floral heart-shaped. Escaped from cultivation.

Eu-phor'-bi-a com-mu-ta'-ta Englm.   **Spurge.** — Stems branched from a commonly decumbent base, 6–12 in. high; leaves obovate, the upper all sessile, the floral ones roundish, dilated, broader than long, umbel 3-forked.

XLIV.   Order **CAL-LIT-RI-CHA'-CE-Æ.   WATER STARWORT FAMILY.** — Aquatic plants; leaves opposite; flowers axillary, monœcious, destitute of proper floral envelopes.

Dicotyls or Exogenous Plants.

1. Genus **CAL-LIT'-RI-CHE**. — Flowers monœcious, solitary or 2 or 3 in the axils, wholly naked; plants low, slender, and usually tufted; floating leaves obovate and 3-nerved, the submersed ones linear.

Cal-lit'-ri-che pa-lus'-tris L. (*C. verna* L.) **Water Starwort.** — Fruit ½ line long, higher than broad; stigmas shorter than the fruit.

Cal-lit'-ri-che het-er-o-phyl'-la Ph. **Water Starwort.** — Fruit smaller, as broad as or broader than high; stigmas as long as the fruit.

XLV. **LIM-NAN-THA'-CE-Æ. FALSE MERMAID FAMILY.** — Low, tender annuals; flowers 3-merous; sepals persistent; carpels nearly distinct, with a common style, 1-ovuled; leaves pinnate.

1. Genus **FLŒR'-KE-A.** — Sepals 3, petals 3, shorter than the calyx, stamens 6, ovaries 3; flowers minute, solitary on axillary peduncles.

Flœr'-ke-a pro-ser-pin-a-coi'-des Willd. **False Mermaid.** — Leaflets 3–5, lanceolate, sometimes 2–3-cleft; a small plant in marshes and on riverbanks.

XLVI. Order **AN-A-CAR-DI-A'-CE-Æ. CASHEW FAMILY.** — Trees or shrubs; flowers small, regular; ovary 1-celled, but styles or stigmas 3.

1. Genus **RHUS.** — Trees or shrubs with resinous or milky, acrid juice; flowers small, regular; juice or exhalations often poisonous.

A. *Flowers in a terminal, thyrsoid panicle; fruit crimson.*

Rhus hir'-ta (L.) Sudw. (*R. typhina* L.) **Staghorn Sumach.** — Branches and stalks *densely-velvety hairy*; leaflets 11–31, pale beneath.

Rhus glab'-ra L. **Smooth Sumach.** — *Smooth*, somewhat glaucous; leaflets 11–31, whitened beneath.

Rhus co-pal-li'-na L. **Dwarf Sumach.** — Branches and stalks downy; petioles *wing-margined* between the 9–21 leaflets; flowering later than the preceding one.

A. *Flowers in loose, slender, axillary panicles; fruit whitish.*

Rhus ver'-nix L. (*R. venenata* DC.) **Poison Elder, Sumach** or **Dogwood.** — Shrub 6–18 ft. high; leaflets 7–13, entire; the most poisonous species; growing in swamps.

Spring Flora of Ohio.

Rhus rad'-i-cans L. (*R. toxicodendron* L.) **Poison Ivy: Poison Oak.—** Climbing by rootlets or sometimes low and erect; leaflets 3, variously notched, sinuate or cut-lobed; poisonous.

    A. *Flowers in solitary or clustered spikes or heads in early spring.*

Rhus ar-o-mat'-i-ca Ait. (*R. canadensis* Marsh.) **Aromatic Sumach.—** A straggling bush, 3-7 ft. high; leaflets 3, unequally cut-toothed.

XLVII. Order **AQ-UI-FO-LI-A'-CE-Æ.** **HOLLY FAMILY.—** Trees or shrubs; flowers axillary, greenish, 4 to 8-merous; calyx minute, stamens on base of corolla.

    Petals oval or obovate, pedicels mostly clustered........................*Ilex.* 1
    Petals linear, pedicels solitary...*Ilicioides.* 2

1. Genus **I'-LEX.—** Flowers more or less diœciously polygamous; petals oval or obovate; fruit a berry-like drupe, containing 4-6 small nutlets.

I'-lex o-pa'-ca Ait. **American Holly.** — Small tree with deep-green, somewhat glossy leaves, evergreen, armed with spiny teeth; fruit red.

I'-lex ver-ti-cil-la'-ta (L.) Gr. **Black Alder: Winterberry.—** Shrub with oval, obovate or wedge-lanceolate, pointed leaves, acute at base, serrate, not evergreen; fruit red.

2. Genus **I-LI-CI-OI'-DES.** (*Nemopanthes.*) — Flowers polygamous, diœcious; petals oblong-linear; drupe with 4-5 bony nutlets.

I-li-ci-oi'-des mu-cro-na'-ta (L.) Britt. (*Nemopanthes fascicularis* Raf.) **Mountain Holly.** — Shrub, much branched; leaves oblong, entire or slightly toothed. Damp, cold woods, rare.

XLVIII. Order **CE-LAS-TRA'-CE-Æ.** **STAFF-TREE FAMILY.** —Shrubs; flowers small, regular; stamens 4-5 on a disk in the bottom of the calyx.

    Erect shrubs, leaves opposite..................................*Euonymus.* 1
    Shrubby climber, leaves alternate...............................*Celastrus.* 2

1. Genus **EU-ON'-Y-MUS.—**Flowers perfect; petals rounded; shrub with 4-sided branchlets; flowers small, in loose cymes or axillary pedicels.

Dicotyls or Exogenous Plants.

Eu-on'-y-mus at-ro-pur-pu'-re-us Jacq. **Waahoo: Burning Bush.** — Upright, 6–14 ft. high; leaves petioled, oval oblong.

Eu-on'-y-mus ob-o-va'-tus Nutt. (*E. americanus* var. *obovatus* T. & G.) — Trailing with rooting branches; flowering stems 1–2 ft. high; leaves obovate or oblong.

2. Genus **CE-LAS'-TRUS.** — Flowers polygamo-dicecious; petals crenulate; pod globose, orange-color, opening by 3 valves, displaying the scarlet covering of the seeds.

Ce-las'-trus scan'-dens L. **Climbing Bitter-sweet; Wax-work.** — Leaves ovate-oblong, finely serrate; a twining shrub.

XLIX. Order **STAPH-Y-LE-A'-CE-Æ. BLADDER-NUT FAMILY.** — Shrubs with opposite, pinnate, stipulate leaves; sepals and petals each 5, colored; fruit an inflated pod.

1. Genus **STAPH-Y-LE'-A.** — Calyx deeply 5-parted, the lobes erect, whitish; petals 5, erect, spatulate; pod large, inflated, 3-lobed, 3-celled.

Staph-y-le'-a tri-fo'-li-a L. **Bladder-nut.** – Leaflets 3, ovate, pointed; branches greenish, striped.

L. Order **A-CER-A'-CE-Æ. MAPLE FAMILY.** — Trees or shrubs; flowers small, regular, petals often wanting; ovary 2-lobed, 2-celled; fruit 2-winged.

1. Genus **A'-CER.** — Flowers polygamo-dicecious, calyx colored, petals present or wanting; fruit a double samara; trees or shrubs; leaves opposite, palmataly lobed or pinnate.

A. *Leaves simple, 3-5-lobed.*

 b. *Flowers in racemes, appearing after the leaves; shrubs or small trees.*

A'-cer penn-syl-van'-i-cum L. **Striped Maple.** — Leaves 3-lobed at the apex, finely and sharply doubly-serrate; racemes drooping, loose; petals obovate; fruit with large divergent wings.

A'-cer spi-ca'-tum Lam. **Mountain Maple.** — Leaves downy beneath, 3 (or 5)-lobed, coarsely serrate; racemes upright, dense; petals linear-spatulate; fruit with small or divergent wings.

78   *Spring Flora of Ohio.*

b. *Flowers in umbellate corymbs, appearing with the leaves.*

A'-cer sac-char'-um Marsh. (*A. saccharinum* Wang.) **Sugar or Rock Maple.**—Bark gray; internodes mostly slender and elongated, commonly glossy and reddish; buds gray, conical, slender and acute; no stipules; leaves 5-lobed (a few 3-lobed) with narrow sinuses, 4–7 in. broad, base truncate or slightly cordate with open sinus, light green above, grayish below.

A'-cer sac-char'-um bar-ba'-tum (Mx.) Trel. **Sugar or Rock Maple.**—Bark gray to almost black; internodes often shorter and stouter, commonly dull but reddish; buds gray pubescent or dark, conical ovoid, often obtuse; no stipules; leaves 3-lobed with open sinuses (lateral lobes often with dilatations), usually about 4 in. broad, somewhat glossy above, pale or glaucous (often downy) beneath.

A'-cer ni'-grum Mx. f. (*A. saccharinum* var. *nigrum* T. & G.) **Black Sugar-Maple.**—Bark nearly black; internodes stout, sometimes short, dull, buff; buds dark, ovoid, often obtuse; stipules adnate triangular or oblong, foliaceous; leaves 3–5-lobed, usually 5–6 in. broad, with drooping sides, often cordate with closed sinus, dull and dark green above, clear green and usually downy below.

b. *Flowers in umbel-like clusters, much preceding the leaves.*

A'-cer sac-cha-ri'-num L. (*A. dasycarpum* Ehrh.) **White or Silver Maple.**—Leaves very deeply 5-lobed, with the sinuses rather acute, silvery-white underneath, the divisions narrow, cut-lobed and toothed; petals none; fruit woolly when young, with large divergent ~~teeth~~. *wings*.

A'-cer ru'-brum L. **Red or Swamp Maple.**—Leaves 3–5-lobed with acute sinuses, whitish underneath, the lobes irregularly serrate and notched; petals linear-oblong; fruit smooth, on prolonged, drooping pedicels.

A. *Leaves pinnate; leaflets 3–5.*

A'-cer ne-gun'-do L. (*Negundo aceroides* Moench.) **Box Elder; Ash-leaved Maple.**—Flowers diœcious, petals none, sterile flowers on capillary pedicels, the fertile in drooping racemes; tree, the twigs light-green.

LI. Order **HIP-PO-CAS-TAN-A'-CE-Æ. BUCKEYE FAMILY.**—Flowers mostly unsymmetrical and irregular; leaves opposite, exstipulate, digitate.

Dicotyls or Exogenous Plants.

1. Genus **AES'-CU-LUS.**—Calyx tubular, 5-lobed; petals 4 or 5; stamens 7 (or 6 or 8); fruit a leathery pod; trees or shrubs.

Aes'-cu-lus glab'-ra Willd. **Fetid or Ohio Buckeye.**—Fruit covered with prickles when young; stamens longer than the corolla; leaflets usually 5.

Aes'-cu-lus oc-tan'-dra Marsh. (*Ae. flava* Ait.) **Sweet Buckeye.**—Fruit smooth; stamens included in the corolla; leaflets 5, sometimes 7, glabrous or often minutely downy beneath; bark of trunk whiter than in the preceding.

LII. Order **RHAM-NA'-CE-Æ. BUCKTHORN FAMILY.**—Shrubs or small trees, flowers small, regular, sometimes apetalous, stamens perigynous, inserted opposite the petals.

Petals small or none, drupe berry-like..........................*Rhamnus*. 1
Petals long-clawed, hooded, fruit dry..........................*Ceanothus*. 2

1. Genus **RHAM'-NUS.**—Calyx 4 to 5-cleft; shrubs or small trees, with greenish polygamous or diœcious flowers in axillary clusters.

Rham'-nus al-ni-fo'-li-a L'Her. **Buckthorn.**—Calyx-lobes and stamens 5, petals wanting; leaves oval, serrate; low shrub, in swamps.

Rham'-nus lan-ce-o-la'-ta Ph. **Buckthorn.**—Calyx-lobes, petals and stamens 4; leaves oblong-lanceolate and acute or on flowering shoots oblong and obtuse, finely serrulate; tall shrub.

2. Genus **CE-AN-O'-THUS.**—Calyx 5-lobed, incurved, petals hooded on spreading claws; filaments elongated; fruit 3-lobed, dry.

Ce-an-o'-thus a-mer-i-ca'-nus L. **New Jersey Tea; Red-root.**—Leaves ovate or oblong-ovate, 3-ribbed, serrate; the common peduncles elongated.

Ce-an-o'-thus o-va'-tus Desf. **Red-root.**—Leaves narrowly oval or elliptic-lanceolate, finely glandular-serrate; common peduncles short.

LIII. Order **VI-TA-CE-Æ. VINE FAMILY.**—Shrubs usually climbing by tendrils; flowers regular, calyx minute or truncate, stamens opposite the very deciduous petals; leaves palmately veined or compound.

Corolla caducous without expanding; leaves simple..................*Vitis*. 1
Corolla expanding; leaves digitate..........................*Parthenocissus*. 2

80   *Spring Flora of Ohio.*

1. Genus **VI'-TIS**. — Flowers polygamo-diœcious, calyx very short, with a nearly entire border or none at all; petals separating only at base; flowers in a compound thyrse, very fragrant.

   A. *A tendril or inflorescence opposite each leaf.*

Vi'-tis la-brus'-ca L. **Northern Fox-Grape.** — Branchlets and young leaves very woolly; leaves large, entire or deeply lobed, slightly dentate, continuing mostly woolly beneath; berries large.

   A. *Tendrils intermittent, none opposite each third leaf.*
      b. *Leaves pubescent and floccose beneath when young.*

Vi'-tis æs-ti-va'-lis Mx. **Summer Grape.** — Leaves large, entire or more or less deeply and obtusely 3 to 5-lobed, very woolly and mostly red or rusty when young.

Vi'-tis bi'-col-or Le Conte. **Summer Grape.** — Differs from the preceding in having the leaves smoothish when old and pale or glaucous beneath.

      (b. *Leaves glabrous or short-hairy, incisely lobed or undivided.*

Vi'-tis cor-di-fo'-li-a Mx. **Frost or Chicken Grape.** — Leaves 3-4 in. wide, not lobed or slightly 3-lobed, cordate with a deep sinus, coarsely and sharply toothed; stipules small.

Vi'-tis ri-pa'-ri-a Mx. **Frost Grape.** — Differing in the larger and more persistent stipules (2-3 lines long); leaves more shining and more usually 3-lobed, with large, acute or acuminate teeth.

2. Genus **PAR-THE-NO-CIS'-SUS**. (*Ampelopsis.*) — Calyx slightly 5-toothed; petals concave, thick, expanding before they fall; leaves digitate, leaflets oblong-lanceolate, sparingly serrate.

Par-the-no-cis'-sus quin-que-fo'-li-a (L.) Planch. (*Ampelopsis quinquefolia* Mx.) **Virginia Creeper: American Ivy; Woodbine: Ampelopsis.** — Woody vine, climbing extensively, clinging by disk-like terminations of the tendrils, and on older parts by aerial rootlets.

Par-the-no-cis'-sus vi-ta'-ce-a (Knerr.) Hitch. **Virginia Creeper.** — Differs from the preceding in having no aerial roots; canes smooth and lighter colored; tendrils usually without disks; flowering and fruiting earlier; fruit larger. Not reported, but doubtless occurring in Ohio.

LIV. Order **TIL-I-A'-CE-Æ**. **LINDEN FAMILY.** — Trees, with mucilaginous and fibrous bark; sepals deciduous; stamens polyadelphous.

1. Genus **TIL'-I-A.** — Sepals 5, petals 5, spatulate-oblong, stamens numerous, filaments cohering in 5 clusters; trees with soft, white wood and tough inner bark; flowers in cymes, peduncles ligulate-winged.

Til'-i-a a-mer-i-ca'-na L. **Basswood.** Leaves large, green and glabrous, or nearly so, thickish; fruit ovoid.

Til'-i-a het-er-o-phyl'-la Vent. **White Basswood.** — Like the preceding, but leaves smooth and bright-green above, silvery-whitened with a fine down underneath; fruit globose.

LV. Order **MAL-VA'-CE-Æ. MALLOW FAMILY.** — Herbs (our species); leaves alternate, stipulate; flowers regular, stamens numerous, monadelphous, their column united to base of petals.

Involucel of 3 bractlets, ovaries united in a ring..................*Malva*. 1
Involucel of many bractlets, pod 5-celled..........................*Hibiscus*. 2

1. Genus **MAL'-VA.** — Styles numerous; petals obcordate; calyx surrounded by 3 bractlets; fruit depressed.

Mal'-va ro-tun-di-fo'-li-a L. **Common Mallow.**—Stem procumbent; leaves round-cordate, on very long petioles; petals twice the length of the calyx.

Mal'-va syl-ves'-tris L. **High Mallow.** — Stem erect, 2-3 ft. high, branched; leaves sharply 5-7-lobed; petals thrice the length of the calyx.

2. Genus **HI-BIS'-CUS.** — Calyx 5-cleft, surrounded by a row of bractlets; styles united; stigmas 5, capitate; fruit a 5-celled pod.

Hi-bis'-cus tri-o'-num L. **Bladder Ketmia.** Low, rather hairy, annual; leaves 3-parted, divisions lanceolate; corolla sulphur-yellow with a blackish eye; fruiting-calyx inflated.

LVI. Order **HY-PER-I-CA'-CE-Æ. ST. JOHN'S-WORT FAMILY.**— Herbs or shrubs; leaves opposite or entire, dotted; flowers regular, stamens numerous, often collected into 3 or more clusters.

1. Genus **HY-PER'-I-CUM.**— Flowers yellow; sepals and petals 5, stamens sometimes in 3 or 5 clusters; styles 3, often more or less united into one; pod 3-celled.

Hy-per'-i-cum pro-lif'-i-cum L. **Shrubby St. John's-wort.** — Branchlets 2-edged; leaves narrowly oblong, 1-2 in. long; shrub 2-6 ft. high; flowers in single or compound clusters; pods 4-6 lines long.

Hy-per'-i-cum den-si-flo'-rum Ph. **St. John's-wort.** Very much branched above, 1-6 ft. high, slender, crowded with smaller leaves; flowers smaller, ½-⅔ in. in diameter, in crowded compound cymes; pods 2-3 lines long; much like the preceding. Very rare.

Hy-per'-i-cum per-fo-ra'-tum L. **St. John's-wort.** — Petals and anthers with black dots; stamens in 3 or 5 clusters; stems much-branched and corymbed, somewhat 2-edged; an introduced perennial herb.

LVII. Order **CIS-TA'-CE-Æ. ROCK-ROSE FAMILY.** — Low shrubs or herbs; flowers regular, stamens indefinite, distinct and hypogynous, calyx persistent.

1. Genus **HE-LI-AN'-THE-MUM.** — Petals 5, yellow, crumpled in the bud, fugacious; flowers of two sorts — the earlier ones with large petals, the later ones smaller.

He-li-an'-the-mum can-a-den'-se (L.) Mx. **Frost-weed.** — Stems at first simple, petal-bearing; flowers solitary; leaves lanceolate-oblong.

LVIII. Order **VI-O-LA'-CE-Æ. VIOLET FAMILY.** — Herbs; corolla irregular, 1-spurred or gibbous, style club-shaped, flower 5-merous; ovary 1-celled, 3 parietal placentas.

Sepals auricled, lower petal spurred... *Viola*. 1
Sepals not auricled, petals nearly equal in length.................... *Solea*. 2

1. Genus **VI'-O-LA.** — Sepals extended into ears at the base; petals somewhat unequal, 1-spurred at the base; two of the stamens spurred and enclosed by the spur of the corolla.

  A. *Stipulus never leaf-like, the lower more or less scarious.*
    b. *Stemless, the leaves and scapes from a rootstock or runners.*
      c. *Stigma large, naked, not beaked.*

Vi'-o-la pe-da'-ta L. **Birdfoot Violet.** — Rootstock erect, not scaly; leaves 3 to 5-divided or the earliest only parted, the lateral divisions 2 to 3-parted, all linear or narrowly-spatulate; flower large.

      c. *Stigma small, naked, often beaked or pointed.*
        d. *Rootstock fleshy and thickened, not filiform; lateral petals bearded.*)

Dicotyls or Exogenous Plants. 83

Vi'-o-la pal-ma'-ta L. **Common Blue Violet.**— Early leaves roundish-cordate or reniform and merely crenate, the sides rolled inward when young, the later ones palmately or pedately lobed or parted, the segments obovate or linear.

Vi'-o-la ob-li'-qua Hill. (*V. palmata* var. *cucullata* Gr.) **Common Blue Violet.**— As the preceding but the later leaves merely crenate, not lobed.

Vi'-o-la sag-it-ta'-ta Ait. **Arrow-leaved Violet.**— Leaves ranging from oblong-heart-shaped to halbred-shaped, arrow-shaped, oblong-lanceolate or ovate, denticulate, sometimes cut-toothed near the base; the lateral (or all) petals bearded.

d. *Rootstocks long and filiform; extensively climbing.*

Vi'-o-la blan'-da Willd. **Sweet White Violet.**— Flowers small, white, short-spurred, petals mostly beardless, the lower strongly veined; leaves round-cordate or kidney-form.

Vi'-o-la lan-ce-o-la'-ta L. **Lance-leaved Violet.**— Leaves lanceolate, erect, blunt, tapering into a long-margined petiole, almost entire; petals beardless, white, small, spur short.

Vi'-o-la ro-tun-di-fo'-li-a Mx. **Round-leaved Violet.**— Flowers yellow, lateral petals bearded and marked with brown lines; leaves round-ovate, cordate, slightly crenate, 1 in. broad, becoming 3-4 in. broad in summer.

b. *Leafy stemmed.*
e. *Stipules entire, spur very short, stems erect.*
f. *Stems naked below; flowers yellow.*

Vi'-o-la pu-bes'-cens Ait. **Downy Yellow Violet.**— Softly pubescent, 6-12 in. high; leaves very broadly heart-shaped; stipules ovate or ovate-lanceolate, large.

Vi'-o-la scab-ri-us'-cu-la (T. & G.) Schw. (*V. pubescens* var. *scabriuscula* T. & G.) **Yellow Violet.**— Differs from the last in being smaller, 4-10 in. high, greener, and slightly pubescent.

Vi'-o-la has-ta'-ta Mx. **Halbred-leaved Violet.**— Nearly glabrous, slender, 4-10 in. high; stem-leaves halbred-shaped or oblong-cordate, slightly serrate, acute; stipules ovate, small. Northern Ohio.

f. Stems more leafy, flowers white or purplish.

Vi'-o-la can-a-den'-sis L. **Canada Violet.** Upright, 1-2 ft. high; leaves cordate, pointed serrate; stipules ovate-lanceolate; the lateral petals bearded.

e. Stipules fringe-toothed, stems erect or spreading.

Vi'-o-la stri-a'-ta Ait. **Pale Violet.** — Stems angular, ascending, 6-10 in. high; leaves cordate, finely serrate; stipules oblong-lanceolate; spur thickish, much shorter than the cream-colored or white petals, the lateral ones bearded.

Vi'-o-la ros-tra-ta Ph. **Long-spurred Violet.** - Stems ascending, 3-6 in. high; leaves roundish-cordate, serrate; stipules lanceolate, large; spur slender ½ in. long, longer than the pale violet, beardless petals.

Vi'-o-la lab-ra-dor'-i-ca Schrank. (*V. canina* var. *muhlenbergii* Gr.) **Dog Violet.** Low (3-7 in.), mostly glabrous; leaves cordate or kidney-form, crenate; stipules lanceolate; spur cylindrical, half the length of the light violet petals, the lateral ones slightly bearded.

A. *Stipules large, leaf-like and lyrate-pinnatifid.*

Vi'-o-la tri'-col-or L. **Pansy: Heart's-ease.** — Petals variable in color, large. Introduced.

Vi'-o-la te-nel'-la Muhl. (*V. tricolor* var. *arvensis* Gr.) Petals shorter, or little longer than the calyx, yellowish-blue, spotted with purple.

2. Genus **SO'-LE-A.** Sepals not prolonged at the base; the lower petal larger and gibbous or saccate at the base; stamens completely united into a sheath enclosing the ovary.

So'-le-a con'-col-or Ging. **Green Violet.** — Plant 1-2 ft. high, leafy to the top; 1-3 small greenish flowers in the axils, on short recurved pedicels.

LIX. Order **PAS SI-FLO-RA'-CE-Æ. PASSION FLOWER FAMILY.** - Herbaceous, climbing by tendrils; flowers perfect, stamens 5, mona-delphous; ovary stalked, 1-celled, placentas 3 or 4.

1. Genus **PAS-SI-FLO'-RA.** - Sepals 5, crowned with a fringe; petals 5, on the throat of the calyx; stamens 5, united in a tube which sheathes the long stalk of the ovary; plant climbing by tendrils.

Dicotyls or Exogenous Plants.

Pas-si-flo'-ra lu'-te-a L. **Passion-flower.** Leaves obtusely 3-lobed at the summit, the lobes entire; flowers greenish-yellow.

LX. Order **THY-ME-LÆ-A'-CE-Æ. MEZEREUM FAMILY.** Shrubs with acrid and very tough (not aromatic) bark; flowers perfect, ovary superior, 1-celled.

1. Genus **DIR'-CA.** — Shrub, much branched; leaves oval-obovate, alternate; flowers light yellow, preceding the leaves, 3 or 4 in a cluster; calyx tubular, stamens and style exserted.

Dir'-ca pa-lus'-tris L. **Leatherwood: Moosewood.**—The wood white and soft, but the fibrous bank exceedingly tough. Damp, rich woods.

LXI. Order **EL-Æ-AG-NA'-CE-Æ. OLEASTER FAMILY.**—Shrubs with silvery-scurfy opposite leaves, and small nearly sessile axillary flowers.

1. Genus **LE-PAR-GY-RÆ'-A.** (*Shepherdia.*) — Flowers diœcious, calyx 4-cleft, becoming berry-like in fruit.

Le-par-gy-ræ'-a can-a-den'-sis (L.) Greene. (*Shepherdia canadensis* Nutt.) **Shepherdia.** — Leaves elliptical or ovate, silvery-downy and scurfy with rusty scales beneath.

LXII. Order **LY-THRA'-CE-Æ. LOOSE-STRIFE FAMILY.**- -Herbs; the calyx enclosing (but free from) the ovary and membranous capsule; flowers often dimorphic.

1. Genus **LY'-THRUM.**—Slender herbs; leaves opposite or scattered; calyx cylindrical, striate, 5–7-toothed with little processes in the sinuses, petals 5–7.

Ly'-thrum a-la'-tum Ph. **Loose-strife.**— Branches with margined angles; petals deep purple.

LXIII. Order **A-RA-LI-A'-CÆ. GINSENG FAMILY.**— Herbs or shrubs; flowers in umbels, styles more than 2; fruit a drupe.

Flowers monœciously polygamous or perfect, styles 5, fruit black or purple..*Aralia.* 1
Flowers diœciously polygamous, styles 2 or 3, fruit bright red or yellowish ....*Panax.* 2

1. Genus **A-RA'-LI-A.**—Umbels usually in corymbs or panicles; flowers white or greenish; styles or cells of the fruit, stamens and petals each 5.

A-ra'-li-a spi-no'-sa L. **Hercules' Club; Angelica Tree.**—Shrub or a low tree; the stout stem and stalks prickly; umbels in a large compound panicle.

A-ra'-li-a his'-pi-da Vent. **Bristly Sarsaparilla: Wild Elder.**—Stem bristly, leafy, 1-2 ft. high, terminating in a peduncle bearing several umbels.

A-ra'-li-a nu-di-cau'-lis L. **Wild Sarsaparilla.** Stem scarcely rising out of the ground, smooth, bearing a single, long-stalked leaf (1 ft. high) and a shorter naked scape with 2-7 umbels.

2. Genus **PA'-NAX.**—Stem simple, low, bearing a whorl of 3 palmately 3 to 7-foliolate leaves and a simple umbel on a slender peduncle.

Pa'-nax quin-que-fo'-li-a L. (*Aralia quinquefolia* Dec. & Pl.) **Ginseng.**—Root large and spindle-shaped, often forked, 4-9 in. long, aromatic; leaflets long-stalked, mostly 5; styles mostly 2; fruit bright red.

Pa'-nax tri-fo'-li-a L. (*Aralia trifolia* Dec. & Pl.) **Dwarf Ginseng; Ground-Nut.** Root or tuber globular, deep in the ground, not aromatic; leaflets 3-5, sessile; styles mostly 3; fruit yellowish.

LXIV. Order **UM-BEL-LIF'-ER-Æ. PARSLEY FAMILY.**—Herbs; flowers in umbels; ovary inferior, styles 2; fruit of 2 seed-like, dry carpels (mericarps.)

    Flowers white or greenish a .
    Flowers yellow c .
a. Stems 4-8 ft. high; stout, coarse plants b .
a. Stems very low, or 1-3 ft. high d .
    b. Glabrous, leaflets mucronate-serrate............................*Angelica*. 1
    b. Woolly, leaflets irregularly cut-toothed, outer flowers usually
        larger radiant and 2-cleft.........................*Heracleum*. 2
c. Leaflets serrate, crenate or toothed, lowest leaves sometimes simple;
    fruit glabrous or nearly so; ovoid to oblong; involucels 3-
    leaved ....*Thaspium*. 3
c. Leaves *palmate*, fruit covered with hooked prickled................*Sanicula*. 4
c. Leaflets *entire*, plant glaucous, fruit glabrous...................*Pimpinella*. 5
c. Leaflets sharply serrate or crenately toothed; fruit ovate to oblong;
    involucels of small bractlets e .

Dicotyls or Exogenous Plants.

    d. Leaflets pinnatifid; plant decumbent or assurgent; fruit narrowly oblong, glabrous......................*Chærophyllum*, 6
    d. Leaflets variously toothed; plant 1-3 ft. high; fruit linear or linear-oblong, attenuate at base, with bristly ribs......*Osmorrhiza*, 7
    d. Plants low, nearly acaulescent, flowers few; fruit nearly orbicular e.
  e. Plants 1-3 ft. high, flowers yellow..................*Zizia*, 8
  e. Plants nearly acaulescent, flowers few, white............*Erigenia*, 9

1. Genus **AN-GEL'-I-CA.**—Calyx-teeth obsolete; fruit flattened dorsally, having a double-winged margin; leaves ternately or pinnately compound; umbels large; flowers white or greenish.

    An-gel'-i-ca at-ro-pur-pu'-re-a L. **Angelica.**—Stem dark-purple; leaflets 1-1½ in. broad, sharply mucronate-serrate.

2. Genus **HER-A-CLE'-UM.**—Calyx-teeth minute; fruit broadly-oval or obovate, leaves ternately compound; umbels broad; flowers white; petals obcordate, the outer ones commonly larger and 2-cleft.

    Her-a-cle'-um la-na'-tum Mx. **Cow Parsnip.**—Woolly stem, 4-8 ft. high; leaflets broad.

3. Genus **THAS'-PI-UM.** Calyx-teeth conspicuous; fruit ovoid to oblong; stems 2-5 ft. high; leaves ternately-divided (or the lower simple); flowers yellow.

    Thas'-pi-um tri-fo-li-a'-tum au'-re-um (Nutt.) Britt. (*Th. aureum* var. *trifoliatum* Coult. & Rose.) **Meadow Parsnip.**—Glabrous; root-leaves mostly cordate, serrate or crenately toothed, stem leaves simply ternate (or rarely biternate); flowers deep yellow.

    Thas'-pi-um bar-bi-no'-de (Mx.) Nutt. **Meadow Parsnip.**—Pubescent on the joints; leaves 1-3-ternate; flowers light yellow.

    Thas'-pi-um bar-bi-no'-de an-gus-ti-fo'-li-um Coult. & Rose. **Meadow Parsnip.**—Like the last but narrower and more sharply cut leaflets, and fruit more or less puberulent.

4. Genus **SA-NIC'-U-LA.** — Calyx-teeth manifest; fruit thickly clothed with hooked prickles; leaves palmately-lobed or parted; flowers greenish or yellowish.

    Sa-nic'-u-la mar-y-lan'-di-ca L. **Sanicle; Black Snake-root.**—Sterile flowers many and long pedicelled; the styles longer than the prickles.

Sa-nic'-u-la can-a-den'-sis L. **Sanicle: Black Snake-root.** — Sterile flowers comparatively few and short-pedicelled; the styles shorter than the fruit.

5. Genus **PIM-PI-NEL'-LA.** — Calyx-teeth obsolete; fruit oblong to ovate; leaves 2 to 3-ternate, leaflets entire; flowers yellow.

Pim-pi-nel'-la in-te-ger'-ri-ma Benth & Hook. **Golden Alexanders.** — Glaucous, 1–3 ft. high; flowers yellow.

6. Genus **CHÆ-RO-PHYL'-LUM.** — Calyx-teeth obsolete, fruit narrowly oblong; leaves ternately decompound, pinnatifid leaflets with oblong, obtuse lobes; flowers white; annual, in moist ground.

Chæ-ro-phyl'-lum pro-cum'-bens (L.) Crantz. **Chervil.** — More or less hairy, stem slender, spreading; flowers white; umbel few-rayed.

7. Genus **OS-MOR-RHI'-ZA.** — Calyx-teeth obsolete, fruit linear to linear-oblong with caudate attenuation at base, very bristly; thick aromatic roots; leaves ternately compound; flowers white.

Os-mor-rhi'-za clay-to'-ni (Mx.) B. S. P. (*O. brevistylis* DC.) **Sweet Cicely.** — Villous-pubescent, styles short, $\frac{1}{2}$ in. long.

Os-mor-rhi'-za lon-gis'-ty-lis DC. **Sweet Cicely.** — Glabrous or slightly pubescent; styles longer, 1 line long or more.

8. Genus **ZIZ'-I-A.** — Calyx-teeth prominent; fruit ovate to oblong, flowers yellow, the central fruit of each umbel sessile.

Ziz'-i-a au'-re-a (L.) Koch. **Meadow Parsnip.** — Leaves (except the uppermost) 2 to 3-ternate; fruit oblong, about 2 lines long.

Ziz'-i-a cor-da'-ta (Walt.) DC. **Meadow Parsnip.** — Radical leaves mostly long-petioled, cordate or even rounder, crenately toothed, very rarely lobed or divided; stem-leaves simply ternate or quinate, fruit ovate, $1\frac{1}{2}$ in. long.

9. Genus **ER-I GE'-NI-A.** Calyx-teeth obsolete, petals obovate or spatulate; fruit nearly orbicular and laterally flattened; small glabrous vernal plants, a simple stem from a deep round tuber and bearing one or two 2-3-ternately divided leaves; flowers few, white.

Er-i-ge'-ni-a bul-bo'-sa (Mx.) Nutt. **Harbinger of Spring.** — Stem 3–9 in. high; leaf segments linear-oblong.

Dicotyls or Exogenous Plants.

LXV. Order **COR-NA'-CE-Æ.** **DOGWOOD FAMILY.**—Trees or shrubs; leaves simple; ovary inferior, style one; calyx-limb minute.

Flowers perfect, 4-parted, leaves mostly opposite.................*Cornus*. 1
Flowers diœciously polygamous, 5-parted, leaves attenuate........*Nyssa*. 2

1. Genus **COR'-NUS.**—Flowers perfect; calyx minutely 4-toothed, petals 4, oblong-spreading, stamens 4; fruit a small drupe; leaves entire, and opposite except in one species; flowers small, in cymes or in close heads surrounded by a corolla-like involucre.

A. *Flowers surrounded by a large 4-leaved corolla-like white involucre.*

Cor'-nus can-a-den'-sis L. **Dwarf Cornel ; Bunch-berry.**—Stems 5-7 in. high, leaves of the involucre ovate, fruit globular; leaves crowded into a whorl, ovate or oval.

Cor'-nus flor'-i-da L. **Dogwood.**—Tree 12-40 ft. high; leaves of the involucre obcordate, 1½ in. long; fruit oval; leaves ovate-pointed.

A. *Flowers white, in flat cymes; no involucre.*
  b. *Pubescence woolly or more or less spreading.*

Cor'-nus cir-ci-na'-ta L'Her. **Round-leaved Cornel or Dogwood.**—Shrub 6-10 ft. high; branches greenish, warty-dotted; leaves round-oval, abruptly pointed, woolly beneath, 2-5 in. broad; fruit light blue.

Cor'-nus a-mo'-num Willd. (*C. sericea* L.) **Silky Cornel ; Kinnikinnik.** — Shrub 3-10 ft. high; branches purplish; the branchlets, stalks and lower surface of the narrowly ovate or elliptical-pointed leaves silky-downy (often rusty), pale and dull; fruit pale blue.

Cor'-nus as-per-i-fo'-li-a Mx.—Branches brownish, the branchlets, etc. rough, pubescent; leaves oblong or ovate, on short petioles, pointed, *rough* with a harsh pubescence *above*, and downy underneath; fruit white.

  b. *Pubescence closely appressed, straight or silky or none.*

Cor'-nus sto-len-if'-er-a Mx. **Red-osier ; Dogwood.**—Branches, especially the osier-like shoots of the season, bright red-purple, smooth; leaves ovate, rounded at base, abruptly short pointed, roughish with a minute close pubescence on both sides, whitish underneath.

Cor'-nus can-di-dis'-si-ma Marsh. (*C. paniculata* L'Her.) **Panicled Cornel.**—Shrub 4-8 ft. high, much branched; branches gray, smooth, leaves lanceolate, taper-pointed, acute at base, whitish beneath but not downy; cymes convex, loose, often panicled; fruit white.

Cor'-nus al-ter-ni-fo'-li-a L. f. **Alternate-leaved Cornel.**—Shrub or tree, 8-25 ft. high; branches greenish, streaked with white; leaves alternate, clustered at the ends, ovate or oval, long pointed, acute at base, whitish and minutely pubescent beneath; fruit deep blue, on reddish stalks.

2. Genus **NYS'-SA.**—Flowers greenish, appearing with the leaves, clustered or rarely solitary at the summit of axillary peduncles; staminate flowers numerous in a simple or compound dense cluster of fascicles; pistillate flowers solitary or 2-8, sessile in a bracted cluster; fruit an ovoid or oblong drupe.

Nys'-sa a-quat'-i-ca L. (*N. sylvatica* Marsh.) **Sour or Black Gum; Tupelo; Pepperidge.** Middle-sized tree with horizontal branches; leaves oval or obovate, alternate but mostly crowded at the ends of the branchlets.

LXVI. Order **ER-I-CA'-CE-Æ. HEATH FAMILY.**—Shrubs, sometimes herbs; flowers regular or nearly so; petals 4-5; anthers usually appendaged or opening at the apex.

    Calyx free from the ovary a .
    Calyx-tube adherent to the ovary h .
a. Corolla funnel-form or campanulate with spreading lobes b .
a. Corolla between rotate and campanulate, the anthers in depressions or pits c .
a. Corolla urceolate, ovoid, ovate, cylindric or globular, lobes small d .
a. Corolla salver-form, the tube hairy inside; very fragrant f .
a. Corolla of 5, obovate, spreading, distinct petals: leaves rusty-woolly
    beneath...*Ledum.* 1
    b. Stamens 5 rarely more, long exserted, corolla funnel-form....*Azalea.* 2
    b. Stamens 10 rarely fewer, exserted, corolla bell-shaped.*Rhododendron.* 3
  c. Evergreen mostly smooth shrubs, leaves entire or coriaceous............*Kalmia.* 4
    d. Fruit fleshy g .
    d. Fruit dry, capsular e'.
e. Shrubs, low; corolla urceolate; capsule globular...............*Andromeda.* 5
e. Shrubs, low; much branched; corolla cylindric-oblong; capsule depressed ....*Chamædaphne.* 6
e. Tree; corolla ovate, 5-toothed; capsule oblong-pyramidal.......*Oxydendron.* 7
  f. Prostrate or trailing, scarcely shrubby; leaves evergreen ...*Epigæa.* 8
g. Corolla ovate or urn-shaped; flowers in terminal racemes or clusters i .
g. Corolla cylindric-ovoid or slightly urn-shaped; flowers few or single
    in the axils..*Gaultheria.* 9
    h. Petals 4, short, spreading; berries white................*Chiogenes.* 10
    h. Petals 4, narrow, reflexed; berries red 1.
    h. Petals 5 k .
i. Trailing, leaves thick and evergreen, entire...............*Arctostaphylos.* 11
    k. Ovary 10-celled, 10-seeded................................*Gaylussacia.* 12
    k. Ovary 4-5-celled, many-seeded.............................*Vaccinium.* 13
l. Stems very slender; leaves small, evergreen................*Schollera.* 14

Dicotyls or Exogenous Plants. 91

1. Genus **LE'-DUM**.—Shrubs, 1–3 ft. high; leaves persistent, oblong or linear-oblong, with rusty wool beneath, entire, margins revolute; petals distinct, spreading; flowers white, small, in terminal, umbel-like clusters.

Le'-dum grœn-lan'-di-cum Oedr. (*L. latifolium* Ait.) **Labrador Tea.**—In cold bogs and mountain woods.

2. Genus **A-ZA'-LE-A**.—Calyx mostly small or minute; the 5 (sometimes 10) stamens and the style more or less exserted and declined; leaves deciduous, glandular-mucronate.

A-za'-le-a vis-co'-sa L. (*Rhododendron viscosum* Torr.) **Clammy Azalea: White Swamp-Honeysuckle.**—Leaves oblong-obovate, the margins and midrib, also the branchlets bristly; corolla clammy, the tube much longer than the lobes; flowers appearing after the leaves.

A-za'-le-a nu-di-flo'-ra L. (*Rhododendron nudiflorum* Torr.)—**Purple Azalea: Pinxter-flower.**—Leaves downy underneath; tube of the corolla scarcely longer than the ample lobes, slightly glandular; flowers from flesh color to pink and purple, appearing before or with the leaves.

A-za'-le-a lu'-te-a L. (*Rhododendron calendulaceum* Torr.) **Flame-colored Azalea.** Leaves hairy; tube of the corolla shorter than the lobes, hairy; flowers large, orange usually turning to flame color, appearing with the leaves.

3. Genus **RHO-DO-DEN'-DRON**.—Calyx mostly small or minute, the 10 stamens rarely exserted; leaves coriaceous and persistent.

Rho-do-den'-dron max'-i-mum L. **Great Laurel.**—Leaves 4–10 in. long, very thick, smooth, with somewhat revolute margins; corolla bell-shaped, 1 in. broad, pale rose-color, or nearly white; blooming in July.

4. Genus **KAL'-MI-A**.—Calyx 5-parted; corolla with 10 depressions in which the 10 anthers are lodged; leaves evergreen, coriaceous.

Kal'-mi-a lat-i-fo'-li-a L. **Mountain Laurel: Calico Bush: Spoonwood.**—Leaves mostly alternate, bright green both sides, ovate-lanceolate, or oblong; flowers rose-color or nearly white, in terminal corymbs.

Kal'-mi-a an-gus-ti-fo'-li-a L. **Sheep Laurel: Lambkill: Wicky.**—Leaves commonly opposite or in threes, pale or whitish underneath, light-green above, narrowly-oblong; flowers more crimson and much smaller than in the last; corymbs lateral.

5. Genus **AN-DROM'-E-DA.** Calyx of 5 nearly distinct sepals; corolla urceolate, 5-toothed, 10 stamens, anthers fixed near the middle, opening by a terminal pore; capsule globular.

An-drom'-e-da po-li-fo'-li-a L. **Wild Rosemary.**—Glabrous, 6-8 ft. high; leaves thick and evergreen, linear to lanceolate-oblong, strongly revolute, white beneath.

6. Genus **CHAM-Æ-DAPH'-NE.** (*Cassandra.*)—Calyx of 5, distinct rigid, ovate and acute sepals; corolla cylindrical-oblong, 5-toothed; anther-cells tapering into a tubular beak; low and much-branched shrubs; leaves coriaceous, scurfy, especially beneath; flowers white, in the axils of the upper small leaves.

Cham-æ-daph'-ne ca-lyc-u-la'-ta (L.) Moench. (*Cassandra calyculata* Don.) **Leather Leaf.**—Leaves oblong-obtuse; flowers white, in the axils of the upper small leaves.

7. Genus **OX-Y-DEN'-DRON.**—Calyx of 5 almost distinct sepals; corolla ovate, 5-toothed, puberulent, stamens 10, linear, tapering upward; capsules oblong-pyrimidal; flowers white, in long one-sided racemes; leaves like the peach.

Ox-y-den'-dron ar-bo'-re-um (L.) DC. **Sorrel Tree; Sourwood.**—Tree, 15-40 ft. high.

8. Genus **EP-I-GÆ'-A.**—Sepals ovate-lanceolate; corolla salverform; anthers 10, opening lengthwise; plant prostrate, scarcely shrubby, bristly with rusty hairs; leaves evergreen, rounded and cordate; flowers rose-colored.

Ep-i-gæ'-a re'-pens L. **Ground Laurel: Trailing Arbutus.**—Flowers in early spring, very fragrant.

9. Genus **GAUL-THE'-RI-A.**—Corolla cylindric-ovoid or slightly urn-shaped, 5-toothed; anther-cells each 2-awned at the summit; fruit enclosed by the thickened, fleshy, red calyx, appearing like a globular berry; shrubby or almost herbaceous.

Gaul-the'-ri-a pro-cum'-bens L. **Wintergreen.**—Stems creeping on or below the surface, the flowering branches 3-5 in. high; flowers in the axils, nodding; the bright-red berries and the obovate or oval leaves spicy-aromatic.

Dicotyls or Exogenous Plants.

10. Genus **CHI-OG'-E-NES.** — Calyx-tube adherent to the ovary, the limb 4-parted, persistent; corolla bell-shaped, 4-parted, stamens 8; berry white, globular; plant trailing, creeping, with scarcely woody stems; leaves small, ovate; margin revolute, the lower surface and the branches beset with rigid, rusty bristles.

Chi-og'-e-nes his-pid'-u-la (L.) T. & G. (*C. serpyllifolia* Salisb.) — **Creeping Snowberry.** — Growing in peat-bogs and mossy woods; not common.

11. Genus **ARC-TO-STAPH'-Y-LOS.** — Corolla ovate and urn-shaped, with short, revolute, 5-toothed limb; anthers with 2 reflexed awns on the back; trailing shrubs.

Arc-to-staph'-y-los u'-va-ur'-si (L.) Spreng. — **Bearberry.** — Leaves thick and evergreen, obovate or spatulate, entire, smooth; fruit red.

12. Genus **GAY-LUS-SA'-CI-A.** — Corolla tubular, ovoid or bell-shaped, the border 5-cleft, anther cells tapering upward, opening by a chink at the end; branching shrubs, commonly sprinkled with resinous dots.

Gay-lus-sa'-ci-a du-mo'-sa (Andr.) T. & G. **Dwarf Huckleberry.** — Somewhat hairy and glandular, 1–5 ft. high, bushy; leaves obovate, oblong, mucronate, green both sides; bracts leaf-like, oval; fruit black, insipid.

Gay-lus-sa'-ci-a fron-do'-sa (L.) T. & G. **Blue Tangle; Dangleberry.** — Smooth, 3–6 ft. high, branches slender and divergent; leaves obovate, oblong, blunt, pale, *glaucous beneath;* bracts oblong or linear; fruit dark blue with a white bloom.

Gay-lus-sa'-ci-a res-in-o'-sa (Ait.) T. & G. **Black Huckleberry.** — Much branched, rigid, slightly pubescent when young, 1–3 ft. high; leaves oval, oblong-ovate or oblong, clothed as well as the flowers with shining resinous globules; bracts small, reddish and deciduous; fruit black without bloom. The common Huckleberry.

13. Genus **VAC-CIN'-I-UM.** — Corolla limb 4–5-cleft, revolute; stamens 8 or 10, anthers sometimes 2-awned, the cells separate and prolonged upwards into a tube; shrubs, flowers solitary, clustered or racemed.

A. *Corolla open-campanulate, anthers 2-awned on the back.*

Vac-cin'-i-um stam-i-ne'-um L. **Deerberry: Squaw Huckleberry.** Diffusely branched, 2-3 ft. high; leaves ovate or oval, pale, glaucous or whitish beneath; anthers much exserted.

A. *Corolla cylindraceous to campanulate, 5-toothed; anthers awnless.*
  b. *Plants dwarf or low, ½-2½ ft. high.*

Vac-cin'-i-um penn-syl-van'-i-cum Lam. **Dwarf Blueberry.** — Dwarf, 6-15 in. high, smooth, with green warty stems and branches; leaves lanceolate or oblong, distinctly serrulate with bristle-pointed teeth, smooth and shining both sides.

Vac-cin'-i-um can-a-den'-se Kalm. **Blueberry.** — Low, 1-2 ft. high; leaves oblong-lanceolate or elliptical, entire, downy both sides as well as the crowded branchlets; otherwise like the last.

Vac-cin'-i-um va-cil'-lans Soland. **Low Blueberry.** — Low, 1-2½ ft. high, glabrous, branchlets yellowish-green; leaves obovate or oval, very pale or dull, glaucous (at least underneath), minutely ciliolate-serrulate or entire.

  b. *Plants tall, 5-10 ft. high.*

Vac-cin'-i-um cor-ym-bo'-sum L. **Common or Swamp Blueberry.**—Leaves ovate, oval, oblong or elliptical-lanceolate, with naked entire margins, pubescent or glabrous. In swamps and low thickets.

Vac-cin'-i-um pal'-li-dum Ait. (*V. corymbosum* var. *pallidum* Gr.) **Common or Swamp Blueberry.** — Differs from the preceding in leaves being mostly glabrous, pale or whitish, glaucous especially underneath, serrulate with bristly teeth.

Vac-cin'-i-um at-ro-coc'-cum (Gr.) Heller. (*V. corymbosum* var. *atrococcum* Gr.) **Common or Swamp Blueberry.** — Differing from the preceding in leaves being entire, downy or woolly underneath even when old, as also the branchlets.

14. Genus **SCHOL'-LE-RA.**—Stems very slender, creeping or trailing; leaves small, entire, whitened beneath, evergreen; pedicels erect, the pale rose-colored flower nodding; corolla 4-parted; berries red, acid.

Schol'-le-ra ox-y-coc'-cus (L.) Roth. (*Vaccinium oxycoccus* L.) **Small Cranberry.**—Stems very slender, 4-9 in. long; leaves ovate, acute, with strongly revolute margins, 2-3 lines long.

Dicotyls or Exogenous Plants. 95

Schol'-le-ra ma-cro-car'-pon (Ait.) Britt. (*Vaccinium macrocarpon* Ait.) **Large** or **American Cranberry.**—Stems elongated, 1–4 ft. long; leaves oblong, obtuse, less revolute, 4–6 lines long.

LXVII. Order **PRIM-U-LA'-CE-Æ. PRIMROSE FAMILY.**—Herbs; flowers regular, symmetrical; stamens opposite the petals; ovary 1-celled; style and stigma undivided.

Leaves opposite; stem erect, simple; flowers yellow ....*Naumbergia.* 1
Stem leafy only at the summit; flowers white, star-shaped ......*Trientalis.* 2
Leaves opposite; plant spreading; flowers red, blue or white...*Anagallis.* 3
leaves radical; flo nodding; petals reflexed Dodecatheon 4

1. Genus **NAUM-BER'-GI-A.**—Corolla very deeply 5(or 6–7)-parted into linear divisions, somewhat purplish-dotted, with a small tooth in each sinus; flower-clusters on axillary peduncles.

Naum-ber'-gi-a thyr-si-flo'-ra (L.) Duby. (*Lysimachia thyrsiflora* L). **Tufted Loosestrife.**—Flowers light yellow, in head-like or spike-like clusters, with short peduncles from one or two middle pairs of the leaves.

2. Genus **TRI-EN-TA'-LIS.**—Calyx mostly 7-parted, the divisions linear-lanceolate; corolla mostly 7-parted, spreading; plant low, smooth, with a whorl of thin veiny leaves at the summit.

Tri-en-ta'-lis a-mer-i-ca'-na Ph. **Chickweed Wintergreen.**—Peduncles one or more, bearing a delicate white and star-shaped flower.

3. Genus **AN-A-GAL'-LIS.**—Calyx 5-parted; corolla rotate, 5-parted longer than the calyx; stamens 5, filaments bearded; leaves opposite or whorled, entire; flowers solitary on axillary peduncles; plants low spreading or procumbent.

An-a-gal'-lis ar-ven'-sis L. **Pimpernel.**—Leaves ovate sessile; petals obovate, fringed.
4. Dodecatheon meadia L. Shooting Star. Caly, and corolla reflexed.

LXVIII. **EB-EN-A'-CE-Æ. EBONY FAMILY.**—Tree, leaves alternate, entire; flowers regular, diœcious or polygamous; fruit a berry.

1. Genus **DI-OS-PY'-ROS.**—Calyx and corolla 4–6-lobed; stamens commonly 16 in the sterile and 8 (imperfect) in the fertile flower; tree with alternate entire leaves.

Di-os-py′-ros vir-gin-i-a′-na L. **Persimmon.**—Leaves ovate-oblong, thickish; corolla pale yellow, ½–²⁄₃ in. long in the fertile flowers, much smaller in sterile; fruit plum-like, very astringent, sweet and edible after exposure to frost.

LXIX. Order **O-LE-A′-CE-Æ. OLIVE FAMILY.**—Trees or shrubs; leaves opposite, pinnate, or simple; flowers sometimes apetalous; ovary 2-celled.

Leaves simple, fruit a fleshy drupe or berry a.
Leaves simple; fruit a dry, 2-celled capsule..........................*Syringa.* 1
Leaves pinnate; fruit a winged samara..........................*Fraxinus.* 2
a. Corolla-lobes long, linear; fruit a fleshy drupe...................*Chionanthus.* 3
a. Corolla-lobes short; fruit a berry..........................*Ligustrum.* 4

1. Genus **SYR-IN′-GA.** — Calyx small, 4-toothed; corolla salverform, stamens short, included; leaves simple, entire; shrubs.

Syr-in′-ga vul-ga′-ris L. **Common Lilac.**—Leaves cordate-ovate; flowers showy, lilac, purple or white. Cultivated; occasionally escaped.

2. Genus **FRAX′-I-NUS.** — Calyx small and 4-cleft; corolla absent; fruit a samara, winged at the apex; trees, with pinnate leaves.

A. *Leaflets with short petioles.*

Frax′-i-nus a-mer-i-ca′-na L. **White Ash.**— Branchlets and petioles glabrous; leaflets 7–9, ovate-oblong or lanceolate-oblong, pointed, pale beneath, entire or sparingly serrate or denticulate.

Frax′-i-nus penn-syl-van′-i-ca Marsh. (*P. pubescens* Lam.) **Red Ash.**—Branchlets and petioles velvety-pubescent; leaflets 7–9, ovate or ovate-lanceolate, taper-pointed, almost entire, pale or more or less pubescent beneath.

Frax′-i-nus lan-ce-o-la′-ta Borck. (*F. viridis* Mx.) **Green Ash.**—Glabrous throughout; leaflets 5–9, ovate or oblong-lanceolate, often wedge-shaped at base and serrate above, bright-green both sides.

Frax′-i-nus quad-ran-gu-la′-ta Mx. **Blue Ash.**—Branchlets square at least on vigorous shoots; leaflets 7–9, oblong-ovate or lanceolate, sharply serrate, green both sides.

A. *Lateral leaflets sessile.*

Frax′-i-nus ni′-gra Marsh. (*F. sambucifolia* Lam.) **Black Ash.**—Branches and petioles glabrous; leaflets 7–11, sessile, oblong-lanceolate, serrate, obtuse or rounded at base. In swamps and wet places.

Dicotyls or Exogenous Plants. 97

3. Genus **CHI-O-NAN'-THUS.** — Calyx 4-parted, very small; corolla of 4 long and linear petals; shrub with entire petioled leaves and panicles of showy white flowers.

Chi-o-nan'-thus vir-gin'-i-ca L. **Fringe-tree.** — Cultivated; native farther south.

4. Genus **LI-GUS'-TRUM.** Calyx short-tubular, 4-toothed, deciduous; fruit a 2-celled berry; shrubs with entire leaves and small white flowers, in terminal panicles.

Li-gus'-trum vul-ga'-re L. **Privet** or **Prim.** — Leaves very smooth; berries black. Cultivated; sometimes naturalized.

LXX. Order **GEN-TI-AN-A'-CE-Æ. GENTIAN FAMILY.** — Herbs; flowers perfect and regular; calyx persistent; ovary 1-celled; stamens inserted on the corolla.

Leaves simple, sessile; corolla 4-lobed............................*Obolaria.* 1
Leaves 3-foliate; corolla 5-cleft, bearded inside...............*Menyanthes.* 2

1. Genus **OB-O-LA'-RI-A.** - Calyx of 2, spatulate, spreading sepals; corolla tubular-bellshaped, 4-cleft; a very smooth purplish-green plant, 3-8 in. high, with opposite, obovate leaves.

Ob-o-la'-ri-a vir-gin'-i-ca L. **Obolaria.** — Herbaceous and rather fleshy, the lower leaves scale-like; flowers $1/3$ in. long.

2. Genus **MEN-Y-AN'-THES.** — Calyx 5-parted; corolla short, funnel-form, 5-cleft, the upper surface white-bearded; flowers racemed on the naked scape; leaflets 3, oval or oblong.

Men-y-an'-thes tri-fo-li-a'-ta L. **Buckbean.** — In bogs.

LXXI. Order **A-POC-Y-NA'-CE-Æ. DOGBANE FAMILY.** — Herbaceous or woody; flowers perfect, regular, 5-merous; carpels 2.

Shrubs, the corolla-throat 5-angled................................*Vinca.* 1
Herbs, erect, corolla campanulate..................................*Apocynum.* 2

1. Genus **VIN'-CA.** — Calyx 5-parted; corolla funnel-form or salverform, the orifice 5-angled; leaves opposite, evergreen; trailing shrubs.

Vin'-ca mi'-nor L. **Periwinkle.** — Flowers solitary, axillary, violet varying to purple or white; stems procumbent, several feet in length.

2. Genus **A-POC'-Y-NUM.**—Calyx very small; corolla campanulate, the lobes short; follicles long, sub-linear; herbs with opposite, entire, mucronate leaves.

A-poc'-y-num an-dro-sæ-mi-fo'-li-um L. **Spreading Dogbane.** Branches divergently forking; leaves ovate, distinctly petioled; cymes loose, spreading; corolla with revolute lobes, the tube much longer than the calyx-lobes. Flowering in June.

A-poc'-y-num can-nab'-i-num L. **Indian Hemp.**—Stem and branches upright or ascending, terminated by erect and close, many-flowered cymes; leaves oval or oblong and even lanceolate, short petioled or sessile; corolla with nearly erect lobes, the tube not longer than the calyx-lobes. Flowering later.

LXXII. Order **POL-E-MO NI-A-CE-Æ. PHLOX FAMILY.**—Flowers perfect, regular; ovary 3-celled; calyx persistent; plants herbaceous.

Corolla salver-form; leaves opposite and entire.................*Phlox.* 1
Corolla open bell-shaped; leaves alternate, pinnate..........*Polemonium.* 2

1. Genus **PHLOX.**—Calyx narrow; corolla salver-form with a long tube; leaves opposite, sessile, entire, flowers cymose. Many of the species are cultivated. Those blooming in the Spring are the following:

Phlox pi-lo'-sa L. **Hairy Phlox.**—Stems slender, 1-1½ ft. high, usually hairy, as are also the lanceolate or linear, 1-4 in. long, leaves; calyx-teeth slender awl-shaped and awn-like; lobes of the pink-purple or rose-red (rarely white) corolla obovate, entire.

Phlox di-var-i-ca'-ta L. **Wild Phlox.** Stems spreading or ascending from a decumbent base, 9-18 in. high; leaves oblong-ovate or lanceolate, or the lower oblong-lanceolate, 1½ in. long; calyx-teeth slender, awl-shaped; lobes of the pale lilac or bluish corolla obcordate or wedge-obovate and notched at the end, or often entire.

Phlox sub-u-la'-ta L. **Ground or Moss Pink.**—Depressed in broad mats, evergreen; leaves awl-shaped, lanceolate or narrowly linear, ¼-½ in. long; corolla pink-purple or rose-color with a darker center (sometimes white), lobes notched, rarely entire.

2. Genus **POL-E-MO'-NI-UM.**—Calyx bell-shaped; corolla open bell-shaped or funnel-form; leaves alternate, pinnate, the upper leaflets sometimes confluent.

Dicotyls or Exogenous Plants.

Pol-e-mo'-ni-um rep'-tans L. **Greek Valerian.**—Stems weak, spreading, 6-10 in. high; leaflets 5-15; corolla light blue, about ½ in. wide.

LXXIII. Order **CON-VOL-VU-LA'-CE-Æ. MORNING GLORY FAMILY.**—Generally twining or trailing; flowers regular and perfect; calyx persistent.

Stigmas 2, linear-filiform to subulate or ovate .... *Convolvulus*. 1
Stigma capitate or 2-3-globose .... *Ipomœa*. 2

1. Genus **CON-VOL'-VU-LUS.**—Corolla funnel-form to campanulate; sometimes 2 leafy bracts at base of calyx; plants twining, erect or prostrate.

Con-vol'-vu-lus spi-tha-mæ'-us L. **Upright Convolvulus.**—Downy; stem low and mostly simple, upright or ascending, 6-12 in. long; calyx enclosed in 2 broad leafy bracts; stigmas oval to oblong; corolla 2 in. long.

Con-vol'-vu-lus se'-pi-um L. **Hedge Bindweed.**—Glabrous or more or less pubescent; stem twining or sometimes trailing extensively; calyx enclosed in 2 broad leafy bracts; stigmas oval to oblong; corolla 1½-2 in. long.

Con-vol'-vu-lus ar-ven'-sis L. **Bindweed.**—Stem procumbent or twining and low; no bracts at or near the base of the calyx; stigmas filiform; leaves ovate-oblong, arrow-shaped; corolla ¾ in. long.

2. Genus **IP-O-MŒ'-A.**—Outer sepals commonly larger; corolla funnel-form or nearly campanulate, contorted in the bud; ovary 2-3-celled or often more in cultivated species.

Ip-o-mœ'-a pur-pu'-re-a Lam. **Common Morning Glory.**—Lobes of stigma 3; sepals long and narrow, bristly-hairy below; corolla funnel-form, 2 in. long, purple, varying to white; cultivated and wild.

Ip-o-mœ'-a pan-du-ra'-ta (L.) Meyer. **Man-of-the-earth: Wild Potato-vine.**—Perennial, smooth or nearly so when old, trailing or sometimes twining; sepals smooth, ovate-oblong, very obtuse; corolla open funnel-form, 3 in. long, white with purple in the tube; stigma 2-lobed or entire.

Ip-o-mœ'-a la-cu-no'-sa L. **Small Morning Glory.**—Annual, rather smooth; stem twining and creeping, slender; sepals lance-oblong, pointed, bristly-ciliate or hairy; corolla white, ½-¾ in. long; stigma 2-lobed or entire.

LXXIV. Order **HY-DRO-PHYL-LA'-CE-Æ.** **WATERLEAF FAMILY.** — Flowers regular, stamens 5, inserted on base of the corolla, alternate; styles 2; plants herbaceous.

Ovary with dilated fleshy placentas enclosing the ovules.....*Hydrophyllum*. 1
Ovary with narrow parietal placentas.................................*Phacelia*. 2

1. Genus **HY-DRO-PHYL'-LUM.** — Corolla bell-shaped, 5-cleft; filaments more or less bearded; anthers linear; capsule ripening 1–4 seeds.

Hy-dro-phyl'-lum vir-gin'-i-cum L. **Waterleaf.** — Leaves pinnately divided, divisions 5–7, ovate-lanceolate or oblong, sharply cut-toothed, the lowest mostly 2-parted, the uppermost confluent.

Hy-dro-phyl'-lum can-a-den'-se L. **Waterleaf.** — Leaves palmately 5 to 7-lobed, rounded, unequally toothed, those from the root sometimes with 2-3 small leaflets; often minute teeth in the sinuses of the calyx.

Hy-dro-phyl'-lum ap-pen-dic-u-la'-tum Mx. **Waterleaf.** — Calyx with a *small reflexed lobe in each sinus;* stem-leaves palmately 5-lobed, rounded, the lobes toothed and pointed, the lowest pinnately divided.

2. Genus **PHA-CE'-LI-A.** — Corolla open bell-shaped, 5-lobed; filaments slender, bearded; anthers ovoid or oblong; capsule 4 to 8-seeded.

Pha-ce'-li-a bi-pin-na-tif'-i-da Mx. **Bipinnate Phacelia.** — Corolla campanulate with narrow folds or appendages within, the lobes entire, bright blue, ½ in. broad; seeds 4; plant upright, 1–2 ft. high. Southern Ohio.

Pha-ce'-li-a pursh'-i-i Buckl. **Miami-mist.** — Corolla almost rotate with *fimbriate lobes* and no appendages within; light blue, varying to white; seeds 4.

Pha-ce'-li-a du'-bi-a (L.) Small (*P. parviflora* Ph.) **Small-flowered Phacelia.** — Corolla open, campanulate, with entire lobes and no appendages, bluish white, $\frac{1}{3}$–$\frac{1}{2}$ in. broad; capsule globular, 6–12 seeded; plant diffusely spreading, 3–8 in. high.

LXXV. Order **BO-RA-GIN-A'-CE-Æ.** **BORAGE FAMILY.** — Mostly scabrous or hispid-hairy herbs; flowers regular, stamens epipetalous; ovary deeply 4-parted.

Dicotyls or Exogenous Plants. 101

The 1 nutlets unarmed. mostly smooth and shining a.
The 1 nutlets armed with barbed prickles............... .*Cynoglossum*. 1
a. Corolla trumpet-shaped with open throat, usually blue....... .....*Mertensia*. 2
a. Corolla short salver-form, mostly blue, the throat crested..... ......*Myosotis*. 3
a. Corolla salver-form to funnel-form, mostly yellow, naked or low crested..*Lithospermum*. 4

1. Genus **CYN-O-GLOS'-SUM.** — Corolla funnel-form, the throat closed with 5 obtuse scales; nutlets oblique, covered with barbed or hooked prickles.

Cyn-o-glos'-sum of-fic-i-na'-le L. **Hound's-tongue.** — Clothed with short, soft hairs, leafy; corolla reddish-purple.

Cyn-o-glos'-sum vir-gin'-i-cum L. **Wild Comfrey.** — Roughish with spreading bristly hairs; stem 2–3 ft. high, few-leaved; stem leaves clasping by a cordate base; corolla pale-blue.

2. Genus **MER-TEN'-SI-A.** — Corolla trumpet-shaped, longer than the calyx, naked or with small appendages in the throat; nutlets fleshy; leaves pale, entire; flowers purplish-blue, rarely white.

Mer-ten'-si-a vir-gin'-i-ca (L.) DC. **Lung-wort: Blue Bells: Virginia Cowslip.** - Corolla 1 in. long, the limb spreading, nearly entire; leaves obovate.

3. Genus **MY-O-SO'-TIS.** — Corolla salver-form, the tube about the length of the 5-toothed or 5-cleft calyx; the throat with 5 small and blunt arching appendages.

A. *Calyx open in fruit, its hairs appressed, none hooked or glandular.*

My-o-so'-tis pa-lus'-tris (L.) Relh. **Forget-me-not.** — Stems ascending, smoothish; leaves rough-pubescent; calyx-lobes much shorter than its tube; limb of corolla 3 or 4 lines broad, sky-blue with a yellow eye.

My-o-so'-tis lax'-a Lehm. **Forget-me-not.** — Stems very slender, decumbent, pubescence all appressed; calyx-lobes as long as its tube; limb of corolla 2 or 3 lines broad, paler blue.

A. *Calyx closing or the lobes erect in fruit, clothed with spreading hairs, some minutely hooked or gland-tipped, corolla small.*

My-o-so'-tis ar-ven'-sis Hoffm. **Forget-me-not.** — Hirsute with spreading hairs; leaves oblong-lanceolate, acutish; racemes naked at the base and stalked; corolla blue, rarely white; pedicels spreading in fruit and longer than the 5-cleft, equal calyx.

My-o-so'-tis ver'-na Nutt. **Forget-me-not.** — Bristly-hirsute; leaves obtuse, linear-oblong, or the lower spatulate-oblong; racemes leafy at the base; corolla very small, white; pedicels in fruit erect, rather shorter than the deeply 5-cleft, unequal, very hispid calyx.

4. Genus **LITH-O-SPER'-MUM.**—Corolla funnel-form or sometimes salver-shaped, the open throat naked or with a more or less evident transverse fold or scale-like appendage opposite each spreading lobe.

> A. *Nutlets rough, gray and dull; throat of the nearly white corolla destitute of any evident folds or appendages.*

Lith-o-sper'-mum ar-ven'-se L. **Corn Gromwell.** — Stems erect, 6-12 in. high; leaves lanceolate or linear; corolla scarcely longer than the calyx.

> A. *Nutlets smooth and shining, white; corolla greenish-white or pale yellow, small, with 5 distinct pubescent scales in the throat.*

Lith-o-sper'-mum of-fic-i-na'-le L. **Common Gromwell.** — Leaves thinnish, broadly lanceolate, acute; corolla exceeding the calyx.

Lith-o-sper'-mum lat-i-fo'-li-um Mx. **Puccoon.** — Leaves ovate and ovate-lanceolate, mostly taper-pointed; the root-leaves large and rounded; corolla shorter than the calyx.

> A. *Nutlets white, smooth and shining; corolla large, salver-form or nearly so, deep orange-yellow, the tube much exceeding the calyx and the throat appendaged.*
>
> b. *Corolla-tube one half to twice longer than the calyx.*

Lith-o-sper'-mum gmel-i'-na (Mx.) Hitch. (*L. hirtum* Lehm.) **Puccoon.** — Hispid with bristly hairs; corolla woolly-bearded at the base inside; flowers distinctly peduncled; fruiting calyx ½ in. long.

Lith-o-sper'-mum ca-nes'-cens (Mx.) Lehm. **Puccoon.** — Softly hairy and more or less hoary; flowers sessile, corolla naked at the base within; fruiting calyx ¼ in. long.

> b. *Corolla tube 2-4 times the length of the calyx, its lobes erose-toothed; but the later flowers small and cleistogamous.*

Lith-o-sper'-mum an-gus-ti-fo'-li-um Mx. **Puccoon.** — Minutely rough-strigose and hoary; leaves linear; the early flowers large and showy, the corolla-tube 8-18 lines long, the later (on more diffusely branching plants) with small and pale corollas.

Dicotyls or Exogenous Plants.

LXXVI. Order **LAB-I-A'-TÆ. MINT FAMILY.**—Ovary deeply 4-parted; stems square; corolla mostly labiate, stamens 2-4; herbs.

Stamens 4 a.
Stamens 2; anthers with a long connective............... ......Salvia. 1
a. The upper inner pair longer; plants creeping and trailing.........Glechoma. 2
a. The stamens, etc., not as above b.
 b. Calyx bilabiate with rounded entire lips, closed in fruit......Scutellaria. 3
 b. Calyx not as above; with 4 or 5 lobes or teeth c.
c. Calyx reticulated-veiny, closed in fruit, upper lip flat................Prunella. 4
c. Calyx thin, inflated, almost equally 4-lobed, open...................Synandra. 5
c. Calyx firmer, about 5-nerved, 5-toothed; corolla-throat dilated........Lamium. 6

1. Genus **SAL'-VI-A.** Calyx 2-lipped; corolla deeply 2-lipped, ringent; stamens 2, on short filaments, jointed with the elongated, transverse connective, the upper end bearing a linear, 1-celled anther; flowers large and showy.

Sal'-vi-a ly-ra'-ta L. **Lyre-leaved Sage.**—Leaves mostly lyrately-lobed or pinnatifid; corolla blue-purple.

2. Genus **GLE-CHO'-MA.** (*Nepeta.*)—Calyx tubular, often incurved, obliquely 5-toothed; corolla dilated in the throat, 2-lipped, the upper lip erect, notched or 2-cleft.

Gle-cho'-ma he-der-a'-ce-a L. (*Nepeta glechoma* Benth.) — **Ground Ivy: Gill-over-the-Ground.**—Leaves round kidney-shaped, crenate; corolla light-blue.

3. Genus **SCU-TEL-LA'-RI-A.**—Calyx bell-shaped in flower, 2-lipped, the lips entire, closed in fruit, the upper with a helmet-like, enlarged appendage on the upper sepal; corolla with an elongated, curved, ascending tube, 2-lipped, the upper lip arched.

Scu-tel-la'-ri-a par'-vu-la Mx. **Skull-cap.**—Subterranean stolens moniliform-tuberiferous; minutely downy, 3-6 in. high, branched and spreading; all but the lower leaves sessile and entire, 6-8 lines long; corolla 2-4 lines long; nutlets wingless.

Scu-tel-la'-ri-a ga-ler-ic-u-la'-ta L. **Skull-cap.**—Subterranean stolens not tuberiferous; smooth or a little downy, erect, 1-2 ft. high, simple; leaves serrate, 1-2 in. long; corolla 8-9 lines long; nutlets wingless.

Scu-tel-la'-ri-a ner-vo'-sa Ph. **Skull-cap.**—Nutlets conspicuously winged, each raised on a slender base; stem 10-20 in. high; lower leaves roundish, the middle ovate, toothed, sub-cordate, 1 in. long; corolla bluish, 4 lines long.

4. Genus **PRU-NEL'-LA.** (*Brunella*.) — Calyx tubular bell-shaped, somewhat 10-nerved and reticulated veiny, flattened on the upper side; upper lip broad and flat, truncate, with 3 short teeth, the lower 2-cleft; corolla 2-lipped, upper lip erect, arched, entire; the filaments 2-toothed at the apex, the lower tooth bearing the anther.

Pru-nel'-la vul-ga'-ris L. **Common Self-heal** or **Heal-all.** Stems nearly simple, flowers in a close spike or head.

5. Genus **SY-NAN'-DRA.** — Calyx bell-shaped, inflated, membranous, irregularly veiny, almost equally 4-toothed; corolla with a long tube, much expanded above and at the throat.

Sy-nan'-dra his-pid'-u-la (Mx.) Britt. (*S. grandiflora* Nutt.) **Synandra.** — Corolla 1½ in. long, yellowish white; leaves broadly ovate, cordate, crenate.

6. Genus **LA'-MI-UM.** Calyx with 5, nearly equal teeth; corolla dilated at the throat; upper lip arched, narrowed at the base; middle lobe of the lower lip notched, contracted as if stalked at the base, the lateral ones small; lowest leaves small and long-petioled, the middle ones cordate and doubly-toothed.

A. *Low; flowers small, purple, in few whorls or heads.*

La'-mi-um am-plex-i-cau'-le L. **Dead Nettle.** — Leaves deeply crenate-toothed or cut, the upper ones clasping.

La'-mi-um pur-pu'-re-um L. **Dead Nettle.** Leaves crenate-toothed, all petioled.

A. *Taller; flowers larger, in several axillary whorls.*

La'-mi-um al'-bum L. **Dead Nettle.** — Corolla white, the tube curved upward, obliquely contracted near the base, a ring of hairs inside.

La'-mi-um mac-u-la'-tum L. **Dead Nettle.** — Like the last but flowers purplish, the ring of hairs transverse instead of oblique.

LXXVII. Order **SO-LA-NA'-CE-Æ. NIGHTSHADE FAMILY.** — Commonly rank-scented, flowers perfect, regular, ovary superior, 2-celled, seeds numerous.

Corolla funnel-form or tubular, anthers opening lengthwise.... *Lycium.* 1
Corolla wheel-shaped, anthers opening at the apex............*Solanum.* 2

Dicotyls or Exogenous Plants.

1. Genus **LY'-CI-UM**. — Shrubby, sometimes spiny plants, with alternate and entire small leaves, and mostly axillary, small flowers.

Ly'-ci-um vul-ga'-re Dun. **Matrimony Vine.** — Shrub with long recurved-drooping branches; corolla short funnel-form, greenish purple.

2. Genus **SO-LA'-NUM**. Corolla wheel-shaped, stamens exserted, filaments very short, anthers converging. A large genus, mostly of warm climates, including the Potato (*S. tuberosum*), the Egg-plant (*S. Melongena*); closely related is the Tomato (*Lycopersicum esculentum*).

So-la'-num dul-ca-ma'-ra L. **Bittersweet.** — Climbing and twining, not prickly; flowers purple or blue in small cymes; leaves ovate-cordate, the upper halberd-shaped or with two lobes or leaflets at base.

So-la'-num car-o-li-nen'-se L. **Horse-Nettle.** Hirsute or roughish-pubescent with 4-8-rayed hairs; prickles mostly numerous, yellowish, stout; leaves oblong or ovate, toothed, lobed or pinnatifid; flowers in racemes, larger.

LXVIII. Order **SCROPH-U-LAR-I-A'-CE-Æ**. **FIGWORT FAMILY.** — Herbaceous; flowers perfect; corolla labiate or nearly regular; stamens didynamous or diandrous; ovary 2-celled.

    Leaves opposite a.
    Leaves alternate b.
a. Corolla 2-cleft, the short tube saccate above..........................*Collinsia*. 1
a. Corolla tubular, sterile stamens about as long as the rest.... ..*Pentstemon*. 2
a. Corolla wheel-shaped, almost regular, 4-parted. ....*Veronica*. 3
  b. Calyx tubular, flattened, colored, concealing the corolla-tube..*Castilleja*. 4
  b. Calyx split in front, oblique, capsule flat....................*Pedicularis*. 5

1. Genus **COL-LIN'-SI-A**. — Corolla declined, deeply 2-lipped, the upper lip 2-cleft, the lower 3-cleft, its middle lobe sac-like, enclosing the 4 declined stamens and style; flowers partly colored, in umbel-like clusters.

Col-lin'-si-a ver'-na Nutt. **Innocence.** — Corolla blue and white, twice the length of the calyx.

2. Genus **PENT-STE'-MON**. — Corolla tubular and more or less inflated or bell-shaped, more or less 2-lipped; flowers mostly showy, thyrsoid or racemose-panicled.

Pent-ste'-mon hir-su'-tus (L.) Willd. (*P. pubescens* Sol.) **Beard-tongue.** — Stem 1-2 ft. high, viscid-pubescent (at least in the inflorescence); thyrse narrow; corolla dull violet or purple (or partly whitish), very moderately dilated, the throat nearly closed by a villous-bearded palate; sterile filament densely bearded.

Pent-ste'-mon pent-ste'-mon (L.) Britt. (*P. lævigatus* Sol.) **Beard-tongue.** — Stem 2-4 ft. high, mostly glabrous, except the inflorescence; thyrse broader; corolla white (commonly tinged with purple), abruptly and broadly inflated, the throat widely open; sterile filament thinly bearded above.

Pent-ste'-mon dig-i-ta'-lis (Sweet) Nutt. (*P. lævigatus* var. *digitalis* Gr.) **Beard-tongue.** — Stem sometimes 5 ft. high; corolla white, larger and more abruptly inflated; otherwise like the last.

3. Genus **VE-RON'-I-CA.** — Corolla 4 (rarely 3 or 5)-parted, wheel-shaped; stamens 2, exserted; stigma single; capsule flattened; leaves mostly opposite.

   A. *Flowers in axillary racemes.*

      b. *Capsule turgid, orbicular.*

Ve-ron'-i-ca an-a-gal'-lis L. **Water Speedwell.** — Leaves sessile, most of them clasping by a heart-shaped base, ovate, lanceolate, acute, serrate or entire.

Ve-ron'-i-ca a-mer-i-ca'-na Schw. **American Brooklime.** — Leaves mostly petioled, ovate or oblong, serrate, thickish, truncate or slightly cordate.

      b. *Capsule strongly flattened.*

Ve-ron'-i-ca scu-tel-la'-ta L. **Marsh Speedwell.** — Smooth, slender and weak; leaves sessile, linear, acute, remotely denticulate; racemes, several, very slender and zigzag; flowers few and scattered; pedicels elongated.

Ve-ron'-i-ca of-fic-i-na'-lis L. **Common Speedwell.** - Pubescent, stem prostrate, rooting at base; leaves short-petioled, obovate, elliptical or wedge-oblong, obtuse, serrate; racemes densely, many-flowered; pedicels shorter than the calyx.

   A. *Flowers in a terminal raceme.*

Ve-ron'-i-ca ser-pyl-li-fo'-li-a L. **Thyme-leaved Speedwell.** — Much branched at the creeping base, nearly smooth; leaves ovate or oblong obscurely crenate, the lowest petioled and rounded; receme loose; corolla whitish or pale blue with deeper stripes.

A. *Flowers appearing to be axillary and solitary, mostly alternate.*

c. *Flowers short-pedicelled ; floral leaves reduced.*

Ve-ron'-i-ca per-e-gri'-na L. **Neckweed: Purslane Speedwell.** — Glandular-puberulent or nearly smooth; lowest leaves oval oblong, toothed, thickish, the others sessile; the upper oblong-linear, longer than the almost sessile whitish flowers; capsule orbicular, slightly notched.

Ve-ron'-i-ca ar-ven'-sis L. **Corn Speedwell.** — Hairy, lower leaves ovate, crenate, the uppermost sessile, lanceolate; capsule inversely heart-shaped.

c. *Flowers long-pedicelled, in axils of ordinary leaves.*

Ve-ron'-i-ca a-gres'-tis L. **Field Speedwell.** — Leaves round or ovate, crenate-toothed; flower small; capsule nearly orbicular, sharply notched, many-ovuled, but 1-2 seeded.

Ve-ron'-i-ca by-zan-ti'-na (Sib & Sm.) B. S. P. (*V. buxbaumii* Ten.) **Speedwell.** — Leaves round or heart-ovate, crenately cut-toothed, $\frac{2}{3}$-1 in. long; flower large, nearly $\frac{1}{2}$ in. wide, blue; capsule obcordate-triangular, broadly notched, 16-24 seeded.

Ve-ron'-i-ca he-der-æ-fo'-li-a L. **Ivy-leaved Speedwell.** — Leaves rounded or heart-shaped, 3 to 7-toothed or lobed; flowers small; capsule turgid, 2-lobed, 2 to 4-seeded.

4. Genus **CAS-TIL-LE'-JA.** — Calyx tubular, flattened, tube of the corolla included in the calyx, its upper lip long and narrow; lower lip short; anther-cells unequal; the floral leaves usually dilated, colored and more showy than the yellow or purplish-spiked flowers.

Cas-til-le'-ja coc-cin'-e-a (L.) Spreng. **Painted Cup.** - Root-leaves clustered, mostly entire, those of the stem incised, the floral 3 to 5-cleft, bright scarlet toward the summit (rarely yellow); corolla pale-yellow.

5. Genus **PE - DIC - U - LA'- RIS.** - Corolla 2-lipped, the upper lip arched, flattened, anther cells equal; flowers rather large, in a spike; leaves chiefly pinnatifid.

Pe-dic-u-la'-ris can-a-den'-sis L. **Lousewort; Wood Betony.** — Upper lip of the dull-greenish-yellow and purplish corolla hooded, incurved, 2-toothed under the apex; capsule flat.

LXXIX. Order **LEN-TIB-U-LA-RI-A'-CE-Æ. BLADDERWORT FAMILY.**
Growing in water or wet soil; calyx and usually corolla bilabiate
ovary 1-celled; placenta central.

1. Genus **U-TRIC-U-LA'-RI-A.**—Aquatic plants, immersed, with capillary dissected leaves, bearing little bladders, rarely with few or no leaves or bladders; flowers on scapes; calyx and corolla 2-lipped.

U-tric-u-la'-ri-a vul-ga'-ris L. **Bladder-wort.** Leaves crowded, 2-3-pinnately parted; capillary, bearing many bladders.

U-tric-u-la'-ri-a in-ter-me'-di-a Hayne. **Bladder-wort.** Leaves crowded, 2-ranked, 4-5 times forked, rigid, the bladders borne on separate leafless branches.

U-tri-u-la'-ri-a cor-nu'-ta Mx. **Bladder-wort.** Growing in peat-bogs or sandy swamps, stem strict; leaves entire, rarely seen.

LXXX. Order **OR-O-BAN-CHA'-CE-Æ. BROOM-RAPE FAMILY.**—
Destitute of green foliage; root parasitic; stamens didynamous ovary 1-celled.

Flowers thyrsoid-spicate, stamens included............... ....*Thalesia.* 1
Flowers densely spicate, stamens exserted................... .....*Conopholis.* 2

1. Genus **THA-LE'-SI-A.** (*Aphyllon.*) Corolla somewhat 2-lipped; stigma broadly 2-lipped or crateriform, plants brownish or whitish; the yellowish or purplish flowers and the naked scapes minutely glandular pubescent.

Tha-le'-si-a u-ni-flo'-ra (L.) Britt. (*Aphyllon uniflorum* Gr.) **One-flowered Cancer-root.**—Scapes 3-5 in. high, 1-flowered.

2. Genus **CO-NOPH'-O-LIS.** Flowers in a thick, chestnut-colored or yellowish, scaly spike, the scales becoming dry and hard; corolla tubular, swollen at base; strongly 2-lipped; stamens protruded.

Co-noph'-o-lis a-mer-i-ca'-na (L. f.) Wallr. **Squaw-root: Cancer-root.**—
Growing in clusters among fallen leaves.

LXXXI. **BIG-NO-NI-A'-CE-Æ. BIGNONIA FAMILY.**—Trees or shrubs; flowers large, showy; corolla more or less labiate; seeds flat, usually winged.

Dicotyls or Exogenous Plants.

Leaves compound, tendril-bearing........................*Bignonia.* 1
Leaves compound, not tendril-bearing..................*Tecoma.* 2
Leaves simple; fertile stamens only 2; trees..........*Catalpa.* 3

1. Genus **BIG-NO'-NI-A.** Corolla somewhat bell-shaped, 5-lobed and rather 2-lipped; stamens 4, often a rudimentary fifth; woody climbers.

Big-no'-ni-a cru-cig'-e-ra L. (*B. capreolata* L.) **Cross-vine.**—Leaves of 2 ovate or oblong leaflets and a branched tendril, often a pair of stipule-like leaves in the axils; corolla orange, 2 in. long.

2. Genus **TEC'-O-MA.** Corolla funnel-form, 5-lobed, a little irregular; stamens 4; woody plants, climbing by aerial roots.

Tec'-o-ma rad'-i-cans (L.) DC. **Trumpet Creeper.** Leaves pinnate; leaflets 9-11; corolla orange and scarlet, 2½-3 in. long.

3. Genus **CA-TAL'-PA.**—Trees; leaves simple, entire; calyx deeply 2-lipped; corolla bell-shaped, the undulate, 5-lobed, spreading border irregular, 2-lipped; capsule very long and slender; seeds winged on each side.

Ca-tal'-pa spe-ci-o'-sa Warder. **Western Catalpa.**—Leaves truncate or subcordate at base, slenderly acuminate, soft-downy beneath, inodorous; flowers larger than in the next; lower lobe of the corolla emarginate or deeply notched; pod terete, 8-20 in. long, 17-20 lines in circumference.

Ca-tal'-pa ca-tal'-pa (L.) Karst. (*C. bignonioides* Walt.) **Southern Catalpa; Indian Bean.**—Leaves cordate, pointed, downy beneath, disagreeable (almost fetid) when touched; flowers showy; corolla 1½ in. long, white, tinged with violet, dotted in the throat with yellow and purple; lower lobe entire; pod 8-10 in. long, 9-12 lines in circumference.

LXXXII. Order **PE-DAL-I-A'-CE-Æ.**—Herbs; leaves mostly opposite, or upper alternate; flowers as in the preceding Order; seeds not winged.

1. Genus **MAR-TYN'-I-A.**—Calyx 5-cleft; corolla gibbous, bell-shaped, 5-lobed and somewhat 2-lipped; the fruit with a woody beak, which splits into 2 hooked horns; low, clammy-pubescent annual,

exhaling a heavy odor; flowers racemed, large, white or purplish, or spotted with yellow and purple.

Mar-tyn'-i-a lou-i-si-a'-na Mill. (*M. proboscidea* Glox.) **Unicorn-plant.** — Leaves heart-shaped, entire or undulate, the upper alternate.

LXXXIII. Order **A-CAN-THA'-CE-Æ. ACANTHUS FAMILY.** — Leaves opposite; flowers labiate or irregular; stamens 2–4; ovary 2-celled.

1. Genus **RU-EL'-LI-A.** — Calyx 5-parted; corolla funnel-form with spreading, ample border; flowers rather large, blue or purple, mostly in axillary clusters.

Ru-el'-li-a cil-i-o'-sa Ph. **Ruellia.** — Hirsute, with soft, whitish hairs; leaves nearly sessile, oval or ovate-oblong, 1–2 in. long; tube of the corolla fully twice the length of the setaceous calyx-lobes.

The other species R. strepens L. nearly or quite glabrous; leaves narrowed at base into a petiole; tube of the corolla slightly exceeding the lanceolate or linear calyx-lobes; blooms later July to September.

LXXXIV. Order **PLAN-TA-GIN-A'-CE-Æ. PLANTAIN FAMILY.** — Mostly acaulescent; leaves radical, ribbed; flowers spicate; corolla scarious; ovary free; style filiform.

1. Genus **PLAN-TA'-GO.** Calyx of 4 persistent sepals, mostly membranaceous margins; corolla salver-form or rotate, withering on the pod; stamens exserted; leaves ribbed.

A. *Flowers all perfect; corolla not closed over the fruit.*
  b. *Bracts short; leaves lanceolate or broad, strongly ribbed.*
    c. *Ribs of the broad leaves rising from the midrib.*

Plan-ta'-go cor-da'-ta Lam. **Cordate Plantain.** — Tall, glabrous; leaves cordate or round-ovate, 3–8 in long, long-petioled.

  c. *Ribs of the leaf free to the contracted base.*

Plan-ta'-go ma'-jor L. **Common Plantain.** — Leaves ovate, oblong-oval or slightly cordate, often toothed; spike dense, obtuse; capsule ovoid, circumscissile near the middle, 8 to 10-seeded; seeds reticulated.

Plan-ta'-go ru-gel'-i-i Dec. **Common Plantain.** — Differing from the last as follows: spikes long and thin, attenuate at the apex; capsules cylindraceous-oblong, circumscissile much below the middle, 4 to 9-seeded; seeds not reticulated.

Dicotyls or Exogenous Plants.

Plan-ta'-go lan-ce-o-la'-ta L. **Rib-grass: Ripple-grass; English Plantain.** - Scapes at length much longer than the lanceolate or lance-oblong leaves; spike dense; seeds 2.

b. *Bracts linear, 2 or 3 times as long as the flowers; leaves narrower.*

Plan-ta'-go ar-is-ta'-ta Mx. (*P. patagonica* var. *aristata* Gr.) **Awned Plantain.** — Loosely hairy or green, or becoming glabrous; bracts narrowly-linear, elongated; spike $1\frac{1}{2}$–4 in. long.

A. *Flowers sub-diœcious or polygamo-cleistogamous; corolla of the fertile flower closed over the maturing capsule forming a kind of beak; small annuals or biennials.*

Plan-ta'-go vir-gin'-i-ca L. **Plantain.**—Hairy or hoary-pubescent, 2–9 in. high; leaves oblong, varying to obovate and spatulate-lanceolate, 3 to 5-nerved; spikes 1–2 in. long.

LXXXV. Order RU-BI-A'-CE-Æ. MADDER FAMILY.

—Leaves simple, entire, opposite or verticillate; ovary inferior; stamens epipetalous.

Leaves opposite; fruit a pod, small plants..................*Houstonia.* 1
Leaves opposite or whorled; fruit a pod; large shrub........*Cephalanthus.* 2
Leaves opposite; fruit a 2-eyed berry, plant creeping.........*Mitchella.* 3
Leaves whorled; fruit twin, of 2 1-seeded carpels; herbs........*Galium.* 4

1. Genus **HOUS-TO'-NI-A.**—Calyx 4-lobed, persistent; corolla salverform or funnel-form; ovary 2-celled, 4–20 seeds in each cell; stipules short, entire, connecting the bases of the leaves; flowers dimorphic.

A. *Small, delicate; peduncle 1-flowered, corolla salver-form.*

Hous-to'-ni-a cæ-ru'-le-a L. **Bluets; Innocence.** — Leaves oblong-spatulate, 3–4 lines long; corolla light-blue, pale-lilac or nearly white, with a yellowish eye.

A. *Flowers in small terminal cymes or clusters; corolla funnel-form.*

Hous-to'-ni-a pur-pu'-re-a L. **Houstonia.**— Pubescent or smooth, 8–15 in. high; leaves varying from roundish-ovate to lanceolate, 3–5 ribbed; calyx-lobes longer than the half-free, globular pod.

Hous-to'-ni-a cil-i-o-la'-ta Gr. (*H. purpurea* var. *ciliolata* Gr.) **Houstonia.** Leaves only $\frac{1}{2}$ in. long, thickish, the cauline oblong-spatulate, the radical oval or oblong, rosulate, hirsute-ciliolate.

Hous-to'-ni-a lon-gi-fo'-li-a Gærtn. (*H. purpurea* var. *longifolia* Gr.) **Houstonia.**— Mostly glabrous, thinner-leaved; leaves oblong-lanceolate to linear, 6-20 lines long, the radical oval or oblong, less rosulate, not ciliate.

Hous-to'-ni-a ten-u-i-fo'-li-a Nutt. (*H. purpurea* var. *tenuifolia* Gr.)
Houstonia. Slender, diffuse, 6-12 in. high, with loose inflorescence and almost filiform branches and peduncles; cauline leaves all linear, hardly over 1 line wide.

2. Genus **CEPH-A-LAN'-THUS.** Calyx-tube inversely pyramidal, the limb 4-toothed; corolla tubular, flowers in spherical, peduncled heads, white; shrubs, with ovate or lanceolate-oblong leaves.

Ceph-a-lan'-thus oc-ci-den-ta'-lis. Button-bush. Growing in swamps and along streams.

3. Genus **MITCH-EL'-LA.** — Flowers in pairs with their ovaries united; calyx 4-toothed; corolla funnel-form, 4-lobed, the lobes densely bearded inside; a smooth, trailing small evergreen herb, with round-ovate and shining, petioled leaves.

Mitch-el'-la re'-pens L. **Partridge-berry.** Flowers white, tinged with purple; the scarlet berries remaining over winter.

4. Genus **GA'-LI-UM.**—Calyx-teeth obsolete, corolla 4(rarely 3)-parted, wheel-shaped; stamens 4 (rarely 3), styles 2; flowers small, cymose, stems square, leaves whorled.

A. *Leaves about 8 in a whorl, fruit with hooked prickles.*

Ga'-li-um a-pa-ri'-ne L. **Cleavers; Goose Grass.** Stem weak and reclining, bristle-prickly backward, hairy at the joints; leaves lanceolate, rough on the margins and midrib, 1-2 in. long.

A. *Leaves in 4s, comparatively large and broad, more or less distinctly 3-nerved; fruit hooked-bristly.*

b. *Flowers dull-purple to yellowish-white; peduncles 3-several-flowered.*

Ga'-li-um pi-lo'-sum Ait. **Hairy Cleavers.** — Hairy, leaves oval, 1 in. long, the lateral nerves obscure; peduncles 2-3-forked, the flowers all pedicelled.

Ga'-li-um cir-cæ'-zans Mx. **Wild Licorice.** Smooth or downy; leaves oval or ovate-oblong, ciliate, 1-1½ in. long; peduncles usually once forked, the branches elongated and divergent.

Dicotyls or Exogenous Plants. 113

Ga'-li-um lan-ce-o-la'-tum Torr. **Wild Licorice.**— Nearly glabrous; leaves (except the lowest) lanceolate or ovate-lanceolate, tapering to the apex, 2 in. long.

 b. *Flowers bright white, numerous in a compact panicle.*

Ga'-li-um bo-re-a'-le L. **Northern Bedstraw.** Smooth, 1-2 ft. high; leaves linear-lanceolate; fruit minutely bristly, sometimes smooth.

 A. *Leaves in 4s-6s, small, 1-nerved, fruit smooth except in the last species.*

  c. *Leaves pointless.*

Ga'-li-um tri'-fi-dum L. **Small Bedstraw.** Stems weak, ascending, 5-20 in. high, mostly roughened backwards on the angles; leaves in whorls of 4-6, linear or oblanceolate; peduncles scattered, 1-7-flowered; corolla-lobes and stamens often only 3.

Ga'-li-um con-cin'-num T. & G. **Cleavers.** Stems low and slender, 6-12 in. high, with minutely roughened angles; leaves all in 6s, linear, slightly pointed, the margins upwardly roughened; peduncles 2-3 times forked, diffusely panicled.

  c. *Leaves cuspidately mucronate or acuminate.*

Ga'-li-um as-pret'-lum Mx. **Rough Bedstraw.** Stems much branched, rough backwards with hooked prickles, leaning on bushes, 3-5 ft. high; leaves in whorls of 6, or 4-5 on the branchlets, oval-lanceolate, with almost prickly margins and midrib.

Ga'-li-um tri-flo'-rum Mx. **Three-flowered Bedstraw.**— Stem 1-3 ft. long, bristly-roughened backward on the angles; leaves elliptical-lanceolate, bristle-pointed, with slightly-roughened margins, 1-2 in. long; peduncles 3-flowered, the flowers pedicelled, greenish.

LXXXVI. Order **CAP-RI-FO-LI-A'-CE-Æ. HONEYSUCKLE FAMILY.**— Mostly shrubby; leaves opposite; ovary inferior; stamens inserted on the tube or base of the corolla.

 Corolla wheel-shaped or urn-shaped, regular a.
 Corolla tubular, often irregular b.
a. Leaves pinnate, fruit berry-like......................................*Sambucus.* 1
a. Leaves simple, fruit a 1-seeded drupe................................*Viburnum* 2
 b. Herbs, flowers axillary..*Triosteum.* 3
 b. Shrubs, erect or climbing c.
 c. Corolla regular, bell-shaped; berry 2-seeded...............*Symphoricarpus.* 4
 c. Corolla more or less irregular, tubular; berry several-seeded.......*Lonicera.* 5
 c. Corolla funnel-form, nearly regular; pod 2-celled, many seeded. ..*Diervilla.* 6

1. Genus **SAM-BU'-CUS.** Calyx-lobes minute or obsolete; corolla open urn-shaped, with a broadly spreading, 5-cleft limb; shrubby plants, with pinnate leaves and small, white flowers in compound cymes.

Sam-bu'-cus can-a-den'-sis L. **Elder.**—Leaflets oblong, the lower often 3-parted; cymes flat; fruit black-purple.

Sam-bu'-cus pu'-bens Mx. (*S. racemosa* L.) **Red-berried Elder.**—Leaflets ovate-lanceolate, downy underneath; cymes panicled, convex or pyramidal; fruit bright-red.

2. Genus **VI-BUR'-NUM.** Calyx 5-toothed; corolla spreading, deeply 5-lobed; flowers white, in flat, compound cymes; shrubs, with simple leaves.

A. *The marginal flowers neutral, corollas greatly enlarged and flat.*

Vi-bur'-num al-ni-fo'-li-um Marsh. (*V. lantanoides* Mx.) **Hobblebush.**—Leaves 4-8 in. across, round-ovate, abruptly pointed, closely serrate; the veins and veinlets beneath and the stalks and branchlets very rusty-scurfy; a straggling shrub.

Vi-bur'-num op'-u-lus L. **Cranberry-tree.** Nearly smooth, upright, 4-10 ft. high; leaves 3-5-ribbed, strongly 3-lobed. The Snow-ball is a cultivated form of this species, the whole cyme consisting of sterile flowers.

A. *No enlarged marginal flowers.*

b. *Leaves 3-ribbed, somewhat 3-lobed, stipules bristle-shaped.*

Vi-bur'-num a-cer-i-fo'-li-um L. **Dockmackie; Arrow-wood.** Shrub 3-6 ft. high; leaves soft-downy beneath; cymes small, slender-peduncled.

b. *Leaves coarsely toothed, pinnately veined, stipules narrowly subulate.*

Vi-bur'-num pu-bes'-cens (Ait.) Ph. **Downy Arrow-wood.**—A low, straggling shrub; leaves ovate or oblong-ovate, the veins and teeth fewer and less conspicuous than in the next, the lower surface, and very short petioles soft-downy, at least when young.

Vi-bur'-num den-ta'-tum L. **Arrow-wood.**—Smooth, 5-15 ft. high, bark ash-colored; leaves broadly ovate, very numerously sharp-toothed and strongly veined.

Dicotyls or Exogenous Plants.

b. *Leaves finely serrate or entire, bright green, stipules none.*

Vi-bur'-num cas-si-noi'-des L. **Withe-rod.**—Shoots scurfy-punctate; leaves opaque or dull, ovate to oblong, margins irregularly crenulate-denticulate or sometimes entire; cymes peduncled, about 5-rayed.

Vi-bur'-num len-ta'-go L. **Sweet Viburnum: Sheep-berry.**—Leaves ovate, strongly pointed, closely serrate, cymes compound, sessile.

Vi-bur'-num pru-ni-fo'-li-um L. **Black Haw.**—Leaves oval, obtuse or slightly pointed, finely serrate, smaller than in the preceding. 1-2 in. long; cymes compound, sessile.

3. Genus **TRI-OS'-TE-UM.**—Calyx lobes linear-lanceolate, leaf-like, persistent; corolla tubular, gibbous at base, scarcely longer than the calyx; coarse hairy perennial herbs, the leaves tapering to the base, and connate around the simple stem.

Tri-os'-te-um per-fo-li-a'-tum L. **Feverwort: Horse-Gentian.**—Softly hairy; leaves oval, abruptly narrowed below, downy beneath, flowers brownish-purple, mostly clustered.

Tri-os'-te-um an-gus-ti-fo'-li-um L. **Feverwort: Horse-Gentian.**—Bristly hairy; leaves lanceolate, tapering to the base; flowers greenish cream-color, mostly single in the axils.

4. Genus **SYM-PHOR-I-CAR'-PUS.**—Calyx-teeth short, persistent; corolla bell-shaped, regularly 4–5-cleft; low and branching upright shrubs with oval and short-petioled leaves.

Sym-phor-i-car'-pus sym-phor-i-car'-pus (L.) Macm. (*S. vulgaris* Mx.) **Indian Currant: Coral-berry.** Flowers in the axils of nearly all the leaves, in short dense clusters; style bearded; fruit red.

Sym-phor-i-car'-pus ra-ce-mo'-sus Mx. **Snowberry.**—Flowers in a loose and somewhat leafy interrupted spike; style glabrous; fruit white.

5. Genus **LON-I-CE'-RA.**—Calyx-teeth very short; corolla tubular or funnel-form, often gibbous at base, irregularly or sub-regularly 5-lobed; leaves entire.

A. *Upright bushy shrubs; peduncles axillary; flowers yellowish.*

Lon-i-ce'-ra cil-i-a'-ta Muhl. **Fly-Honeysuckle.**—Branches straggling, 3–5 ft. high; leaves oblong-ovate, often cordate, petioled, thin, downy beneath; filiform peduncles shorter than the leaves.

Lon-i-ce'-ra cæ-ru'-le-a L. **Mountain Fly-Honeysuckle.** Low, 1-2 ft. high, branches upright; leaves oval, downy when young; peduncles very short.

Lon-i-ce'-ra ob-lon-gi-fo'-li-a (Goldie) Hook. **Swamp Fly-Honeysuckle.** — Shrub, 2-5 ft. high, branches upright; leaves 2-3 in. long, oblong; downy when young, smooth when old; peduncles long and slender.

A. *Twining shrubs with the flowers in sessile, whorled clusters from the axils of the often connate upper leaves.*

b. *Corolla trumpet-shaped, almost regular.*

Lon-i-ce'-ra sem-per-vi'-rens Ait. **Trumpet Honeysuckle.** — Flowers in somewhat distant whorls, scentless, nearly 2 in. long, red outside, yellowish within; leaves oblong, the uppermost pairs connate.

b. *Corolla ringent, the lower lip narrow, the upper 4-lobed.*

c. *Corolla-tube 1 in. long, glabrous inside.*

Lon-i-ce'-ra gra'-ta Ait. **American Woodbine.** — Leaves smooth, glaucous beneath, obovate, the 2 or 3 upper pairs united.

c. *Corolla hairy within, the tube ½ in. long or less.*

Lon-i-ce'-ra hir-su'-ta Eaton. **Hairy Honeysuckle.** — Leaves deep-green above, downy-hairy beneath, dull, broadly-oval, the uppermost united, the lower short-petioled; corolla orange-yellow, clammy-pubescent.

Lon-i-ce'-ra sul-li-van'-ti-i Gr. **Sullivant's Honeysuckle.** — At length much whitened with glaucous bloom, 3-6 ft. high, glabrous; leaves oval or obovate-oblong, 2-4 in. long, sessile and mostly connate on the flowering stems; filaments nearly glabrous.

Lon-i-ce'-ra glau'-ca Hill. **Small-flowered Honeysuckle.** — Glabrous, or lower leaf-surface sometimes puberulent, 3-5 ft. high; leaves oblong, 2-3 in. long, glaucous, but less whitened than the last, the 1-4 upper pairs connate; corolla tube only 3-4 lines long within, and also the style and base of filaments hirsute.

2. Genus **DI-ER-VIL'-LA.** — Calyx-tube tapering at the summit, the lobes slender, persistent; corolla funnel-form, almost regular; leaves ovate or oblong, petioled, serrate.

Di-er-vil'-la di-er-vil'-la (L.) Macm. (*D. trifida* Mœnch.) **Bush Honeysuckle.** — Peduncles cymosely about 3-flowered, terminal or from the upper axils.

Dicotyls or Exogenous Plants. 117

LXXXVII. Order **VA-LE-RI-AN-A'-CE-Æ.** **VALERIAN FAMILY.** Herbs; leaves opposite, exstipulate; corolla somewhat irregular; stamens epipetalous; ovary inferior 3-celled, but 1-seeded.

Limb of calyx consisting of inrolled, plumose bristles..........*Valeriana.* 1
Limb of calyx obsolete or merely toothed......................*Valerianella.* 2

1. Genus **VA-LE-RI-A'-NA.** The inrolled pappus-like plumose-bristles (representing the limb of the calyx) unroll and spread as the seed-like, 1-celled fruit matures; perennial herbs with thickened, strong-scented roots and simple or pinnate leaves.

Va-le-ri-a'-na ed'-u-lis Nutt. **Valerian.** — Stem straight, 1-4 ft. high, few-leaved; leaves commonly minutely and densely ciliate, those of the root spatulate and lanceolate, of the stem pinnately parted into 3-7 long and narrow divisions; flowers in a long and narrow panicle, nearly diœcious; corolla whitish.

Va-le-ri-a'-na syl-vat'-i-ca Banks. **Valerian.** - Root-leaves ovate or oblong, rarely with 2 small lobes; stem-leaves pinnate, with 3-11 oblong-ovate or lanceolate, nearly entire leaflets; cymes at first close, many flowered; corolla ¼ in. long, rose-color or white. Growing northward; reported for Ohio.

Va-le-ri-a'-na pau-ci-fio'-ra Mx. **Valerian.** Root leaves ovate, cordate, toothed, sometimes with 2 small lateral divisions; stem-leaves pinnate, leaflets 3-7, ovate, toothed.

2. Genus **VA-LE-RI-AN-EL'-LA.** —Limb of calyx obsolete; fruit 3-celled, two of the cells empty and sometimes confluent into one, the other 1-seeded.

A. *Corolla bluish.*

Va-le-ri-an-el'-la lo-cus'-ta (L.) Bettke. (*V. olitoria* Poll.) **Corn Salad: Lamb-Lettuce.** Corolla bluish; fruit with a corky mass at the back of the fertile cell; empty cells as large as the fertile, contiguous.

A. *Corolla white.*

Va-le-ri-an-el'-la chen-o po-di-fo'-li-a (Ph.) DC. **Corn Salad; Lamb-Lettuce.** Fertile cell broader than the empty ones, beaked, cross-section of fruit triangular.

Va-le-ri-an-el'-la ra-di-a'-ta (L.) Dufr. **Corn Salad: Lamb-Lettuce.** Fertile cell broad as the empty one, beaked; cross-section of fruit quadrate.

Va-le-ri-an-el'-la wood-si-a'-na pa-tel-la'-ri-a (Sull.) Gr. **Corn Salad: Lamb-Lettuce.**—Fertile cell much the narrowest; fruit saucer-shaped, emarginate at base and apex, winged by the divergent cells.

LXXXVIII. Order **CAM-PAN-U-LA'-CE-Æ. CAMPANULA FAMILY.**—Herbs, leaves simple, alternate; corolla regular, stamens 5, mostly distinct; ovary inferior.

Calyx mostly 5-cleft; filaments broad at base ....*Campanula.* 1
Calyx-tube elongated; filaments hairy, shorter than anthers.....*Legouzia* 2

1. Genus **CAM-PAN'-U-LA.**—Calyx 5-cleft; corolla bell-shaped, stamens 5; filaments broad and membranous at base.

Cam-pan'-u-la ro-tun-di-fo'-li-a L. **Harebell.**—Slender, branching, 5-12 in. high; root leaves round-cordate or ovate, long petioled, early withering away; stem leaves linear or narrowly lanceolate, smooth.

Cam-pan'-u-la a-par-i-noi'-des Ph. **Marsh Campanula.**—Stem simple and slender, weak, 8-20 in. high, somewhat 3-angled, rough backward on the angles as are the edges and midrib of the linear lanceolate leaves.

2. Genus **LE-GOU'-ZI-A.** (*Specularia*).—Calyx 5 (or 3-4)-lobed; corolla wheel-shaped, stamens 5.

Le-gou'-zi-a per'-fo-li-a'-la (L.) Britt. (*Specularia perfoliata* A. DC.) **Venus's Looking-Glass.**—Leaves roundish or ovate, clasping by the cordate base, toothed, flowers sessile, solitary or 2-3 in the axils, blue or purplish.

LXXXIX. Order **COM-POS'-I-TÆ. SUNFLOWER FAMILY.**—Flowers in heads surrounded by a scaly involucre, corolla tubular in the perfect flowers and ligulate in the marginal or ray-flowers; anthers united in a tube; ovary inferior; calyx-limb represented by (*pappus*) a cup, teeth, scales, awns or bristles.

Heads with no ray ligulate-flowers c .
Heads with ray-flowers a .
a. Pappus capillary b .
a. Pappus a short crown or none d .
  b. Ray-flowers white or purple or flesh-color.................... *Erigeron* 1
  b. Ray-flowers yellow f .
c. Heads diœcious; floccose-woolly plants........................ *Antennaria.* 2

d. Rays yellow, disk dull brown.................................*Rudbeckia.* 3
d. Rays white e.
e. Receptacle chaffy; heads small, the rays few......................*Achillea.* 4
e. Receptacle not chaffy except at summit; heads rather large........*Anthemis.* 5
e. Receptacle naked, rays many; heads large....................*Chrysanthemum.* 6
  f. Leaves all radical; heads solitary on scaly scapes...............*Tussilago.* 7
  f. Leaves not all radical; heads corymbed..........................*Senecio.* 8

1. Genus **ER-IG'-ER-ON**. — Heads many-flowered, radiate, mostly flat or hemispherical, the narrow rays very numerous; involucral scales narrow, equal and little imbricated; achenes flattened; pappus a single row of capillary bristles.

  A. *Rays white.*

    Er-ig'-er-on an'-nu-us (L.) Pers.   **Daisy Fleabane.** — Stem beset with spreading hairs; leaves coarsely and sharply toothed, the lowest ovate, the upper ovate-lanceolate.

    Er-ig'-er-on ra-mo'-sus (Walt.) B. S. P. (*E. strigosus* Muhl.) **Daisy Fleabane.** — Roughish, with minute appressed hairs, or almost smooth; leaves entire or nearly so, the lowest oblong or spatulate, the upper lanceolate.

  A. *Rays purplish or flesh-color.*

    Er-ig'-er-on pul-chel'-lus Mx. (*E. bellidifolius* Muhl.) **Robin's Plantain.** — Stem simple, rather naked above, bearing few (1–9) large heads; rays about 50, rather broad, light bluish-purple.

    Er-ig'-er-on phil-a-del'-phi-cus L. **Common Fleabane.** — Stem leafy, corymbed, having several small heads; rays innumerable and very narrow, rose-purple or flesh-color.

2. Genus **AN-TEN-NA'-RI-A**. — Heads many-flowered, diœcious; flowers all tubular; involucre dry and scarious; achenes terete or flattish; leaves entire, heads corymbed, rarely single.

    An-ten-na'-ri-a plan-ta-gin-i-fo'-li-a (L.) Rich. **Everlasting.** — Low, 3–18 in. high, spreading by offsets and runners.

3. Genus **RUD-BECK'-I-A**. — Heads many-flowered, radiate; scales of the involucre leaf-like, in about 2 rows, spreading; chaff short, concave; achenes 4-angular, flat on top.

    Rud-beck'-i-a hir'-ta L. **Cone-flower.** — Very rough and bristly-hairy throughout; heads large; leaves nearly entire.

4. Genus **A-CHIL-LE'-A.** Heads many-flowered, radiate, rays few; involucral scales imbricated, with scarious margins; receptacle chaffy; achenes flattened, oblong.

A-chil-le'-a mil-le-fo'-li-um L. **Yarrow: Milfoil.** Leaves twice pinnately parted, the divisions linear, 3-5-cleft, crowded; corymb of heads compound.

5. Genus **AN'-THE-MIS.**—Heads many-flowered, radiate; involucre hemispherical, of many small imbricated, dry and scarious scales; receptacle conical, without chaff near the margin.

An'-the-mis cot'-u-la L. **May-weed; Dog-fennel.**—Leaves finely 3-pinnately dissected; plants ill-scented.

6. Genus **CHRY-SAN'-THE-MUM.**—Rays numerous; involucre broad and flat, the scales with scarious margins, imbricated; receptacle naked; disk-corollas flattened; achenes striate; pappus none.

Chry-san'-the-mum leu-can'-the-mum L. **Ox-eye Daisy: White-Weed.**- Stem erect nearly simple, naked above, bearing a single large head.

7. Genus **TUS-SI-LA'-GO.**—Heads many-flowered; ray-flowers in several rows, narrowly ligulate; involucre nearly simple; achenes cylindrical-oblong; pappus copious, soft and capillary.

Tus-si-la'-go far'-fa-ra L. **Coltsfoot.**—The scapes scaly, 1-flowered, preceding the rounded-cordate, angled or toothed leaves.

8. Genus **SEN-E'-CI-O.**—Heads many-flowered, flowers yellow, involucre simple or with a few bractlets at the base, receptacle flat, naked; pappus of numerous soft capillary bristles.

A. *Root annual.*

Sen-e'-ci-o lo-ba'-tus Pers. **Butter-weed.**—Leaves somewat fleshy, lyrate or pinnate, the divisions crenate or cut-lobed; rays 6-12.

A. *Root perennial.*

Sen-e'-ci-o au'-re-us L. **Golden Ragwort: Squaw-weed.**—Leaves thin, the radical simple and rounded, the larger ones mostly heart-shaped, crenate-toothed, long-petioled; lower stem-leaves lyrate, upper ones lanceolate, cut-pinnatifid, sessile or partly clasping.

Sen-e′-ci-o ob-o-va′-tus Muhl. (*S. aureus* var. *obovatus* T. & G.)
Golden Ragwort: Squaw-Weed.— Differs from the preceding in having the root-leaves thicker, round-ovate, with a cuneate or truncate base, or the earliest almost sessile in rosulate tufts.

Sen-e′-ci-o bal-sam′-i-tæ Muhl. (*S. aureus* var. *balsamitæ* T. & G.)
Golden Ragwort: Squaw-Weed.— Compared with the preceding less glabrate; root-leaves oblong, spatulate or lanceolate, narrowed to the petiole, serrate, the upper lyrate-pinnatifid; heads rather small and numerous.

XC. Order **CI-CHO-RI-A′-CE-Æ. CHICORY FAMILY.** Flowers in heads surrounded by a scaly involucre; corolla ligulate in all the flowers; anthers, calyx and ovary as in the preceding Order; herbs with milky juice and alternate leaves.

Heads small terminating the naked scapes or branches......... *Adopogon*. 1
Head large and solitary, on a hollow scape............... *Taraxacum*. 2

1. Genus **AD-O-PO′-GON.** (*Krigia*). Heads several-many-flowered, corollas all ligulate; involucral scales several, in about 2 rows, thin; achenes short and truncate; pappus double, the outer of thin chaffy scales, the inner of delicate bristles.

Ad-o-po′-gon car-o-li-ni-a′-num (Walt.) Britt. (*Krigia virginica* Willd). **Dwarf Dandelion.**— Pappus of 5-7 short roundish chaff and 5-7 alternating bristles; stems becoming branched and leafy; earlier leaves rounded and entire, the others narrower and often pinnatifid.

Ad-o-po′-gon vir-gin′-i-cum (L.) Kuntze. (*Krigia amplexicaulis* Nutt.) **Dwarf Dandelion.**— Pappus 10-15 small oblong chaff and 10-15 bristles; stem leaves 1-3, oval or oblong, clasping, mostly entire; the radical ones often toothed rarely pinnatifid.

2. Genus **TA-RAX′-A-CUM.**— Head many-flowered, large, solitary, on a slender, hollow scape; corollas all ligulate; involucre double, the outer of short scales; apex of the achene prolonged into a very slender beak, bearing the copious, soft white bristles.

Ta-rax′-a-cum ta-rax′-a-cum (L.) Karst. (*T. officinale* Web.) **Dandelion.**— Leaves radical, pinnatifid or runcinate; flowers yellow.

KEY TO TREES AND SHRUBS.

Cone-bearing plants, or the so-called "evergreens" 1.
Plants not cone-bearing; usually deciduous 4.
                                      PAGE
1. Leaves needle-shaped, 2–5 in a sheath............................*Pinus.* 17
1. Leaves not in a sheath, very many in a cluster, deciduous...........*Larix.* 17
1. Leaves scale-like and adnate or free, and awl-shaped 2.
1. Leaves not as above, scattered 3.
    2. Leaves 2-ranked, fruit a cone with few scales.................*Thuya.* 18
    2. Leaves not 2-ranked, fruit a berry-like cone..............*Juniperus.* 18
3. Leaves green both sides, 2-ranked.................................*Taxus.* 18
3. Leaves whitened beneath, petioled................................*Tsuga.* 18
    4. Leaves simple 25.
    4. Leaves compound 5.
5. Leaves of 2 leaflets and a branched tendril; tall climber.........*Bignonia.* 109
5. Leaves tri-foliate or pinnately compound 7.
5. Leaves palmately compound 6.
    6. Tree, leaves opposite, fruit large, dry.................*Æsculus.* 79
    6. Vine, leaves alternate, fruit small berries*Vitis.* 80
    6. Half shrub stems biennial, leaflets 3 or 5........ ....*Rubus.* 64
7. Leaves all trifoliate 9.
7. Leaves pinnate a few of them may be trifoliate 11.
7. Leaves bipinnate or decompound (a few may be simply pinnate) 8.
    8. Shrubs or small tree, the umbels in a large panicle..........*Aralia.* 86
    8. Tree, thorny, leaflets small, leaves often pinnate ....*Gleditschia.* 69
    8. Tree not thorny, leaflets large, leaves very large.......*Gymnocladus.* 69
9. A climbing vine, sometimes small and erect; leaves alternate.......*Rhus.* 75
9. A climbing vine; fruit plumose; leaves opposite................*Clematis.* 16
9. Half shrub biennial, trailing or nearly erect; leaflets 3 or 5 .....*Rubus.* 64
9. A straggling bush, leaves aromatic when crushed; leaflets cut-
    toothed ....*Rhus.* 75
9. Strictly upright shrubs 10.
    10. Leaves opposite, fruit a bladdery pod, flowers in drooping
        racemes, white; branches greenish-striped...........*Staphylea.* 77
    10. Leaves alternate, fruit surrounded by a wing; flowers in
        cymes; branches not as above................................*Ptelea.* 73
11. More or less thorny or prickly 12.
11. Not at all thorny or prickly 14.
    12. Shrub; bark, leaves and pods aromatic, leaflets 5–9....*Xanthoxylum.* 73
    12. Bark, etc., not pungent-aromatic 13.
13. Large tree with large mostly-branched thorns, fruit a pod...*Gleditschia.* 69
13. Small trees or shrubs, prickly spines for stipules, fruit a pod....*Robinia.* 71
13. Shrubs with conspicuous adnate stipules, fruit fleshy rose-hip .....*Rosa.* 67
13. Half shrubs biennial stems, fruit a blackberry or raspberry.....*Rubus.* 64

Key to Trees and Shrubs.

```
14. Climbing vines  15 .
14. Trees or upright shrubs  17 .
    15. Leaves opposite  16 .                                           PAGE
    15. Leaves alternate, flowers showy, pod elongated......Kraunhia.    71
16. Leaflets 3-7; tails of fruit plumose...........................Clematis.  46
16. Leaflets 9-11; corolla 2½-3 in. long.............................Tecoma. 109
    17. Leaflets 3-5; twigs light green, fruit a double samara..........Acer.  77
    17. Leaflets 5 to many; twigs, etc., not as above  18 .
18. Leaves 2-4 ft. long; leaflets 21-41; tree, bark smooth..........Ailanthus. 73
18. Leaves not so long, leaflets less numerous  19 .
    19. Pith in transverse plates, leaflets many.....................Juglans.  28
    19. Pith not in transverse plates  20 .
20. Shrubs or small trees  21 .
20. Large forest trees  24 .
    21. Shrub not strictly upright, somewhat climbing by aerial root-
          lets................................................................Tecoma. 109
    21. Strictly upright shrubs or small trees  22 .
22. Flowers in panicles; fruit dry; leaflets entire or sinuate...........Rhus.  75
22. Flowers in flat cymes or panicles; fruit a berry; leaflets serrate or doubly
       serrate  23 .
    23. Leaflets serrate; pith large; bark warty; shrub.........Sambucus. 113
    23. Leaflets mostly doubly-serrate; pith not large; small tree or
          shrub..............................................................Sorbus.  62
24. Leaves alternate; fruit a nut hickory-nut .......................Hicoria.  28
24. Leaves opposite; fruit winged at apex............................Fraxinus.  96
    25. Parasitic shrubs on limbs of trees; leaves evergreen..Phoradendron.  37
    25. Not parasitic; leaves opposite or verticillate  26 .
    25. Not parasitic; leaves alternate  37 .
26. Very low, trailing shrubs  27 .
26. Not as above  28 .
    27. Leaves evergreen, smooth and shining, entirely prostrate;
          introduced.........................................................Vinca.  97
    27. Leaves deciduous; flowering stems, 1-2 ft. high; native....Euonymus.  76
28. Leaves all deeply or slightly lobed  29 .
28. Leaves lobed or wavy-toothed only on young shoots, the others en-
       tire ............................................................Symphoricarpus. 115
28. Leaves entire or serrate, not lobed  30 .
    29. Fruit a double samara, in umbellate clusters or racemes.........Acer.  77
    29. Fruit a 1-seeded drupe, in flat cymes.......................Viburnum. 114
30. Leaves evergreen, coriaceous, sometimes in 3s or alternate........Kalmia.  91
30. Leaves deciduous  31 .
    31. Leaves serrate; shrubs  32 .
    31. Leaves entire; shrubs or trees  33 .
32. Fruit a drupe with soft pulp, in flat cymes.... ................Viburnum. 114
32. Fruit, a 2-valved pod, in upper axils or terminal................Diervilla. 116
32. Fruit crowned with 2-4 diverging styles, in flat cymes....... Hydrangea.  59
32. Fruit a 3-5-lobed pod, seeds in a red aril; peduncles axillary..Euonymus.  76
    33. Tree with very broad leaves; pods elongated................Catalpa. 109
    33. Small tree; leaves ovate; fruit a red drupe..................Cornus.  89
    33. Shrubs, upright, spreading or twining  34 .
```

		PAGE
34. Twining, uppermost leaves connate	*Lonicera*,	115
34. Not as above 35.		
35. Fruit in spherical, peduncled heads, dry and hard	*Cephalanthus*,	112
35. Fruit like the rose-hip; bark and foliage aromatic	*Buettneria*,	49
35. Fruit black, berries in terminal panicles	*Ligustrum*,	97
35. Fruit fleshy, drupes, lateral in drooping panicles	*Chionanthus*,	97
35. Fruit on axillary peduncles, 2 berries, sometimes united	*Lonicera*,	115
35. Fruit red or white globose berries, in close, short axillary spikes or clusters	*Symphoricarpus*,	115
35. Fruit yellowish-red oval berries, single or clustered in the axils	*Lepargyraea*,	85
35. Fruit 1-seeded drupes with soft pulp, compressed stone; in flat cymes	*Viburnum*,	114
35. Fruit fleshy, white, blue or red drupes; stone 2-seeded; in cymes or close heads	*Cornus*,	89
35. Fruit a dry pod, not as above 36.		
36. Pod 3-celled; branchlets often 2-edged; leaves narrow	*Hypericum*,	81
36. Pod 2-celled; leaves cordate-ovate; flowers showy; cultivated	*Syringa*,	96
37. Shrubs or small tree more or less prickly or thorny 38.		
37. Not at all prickly or thorny 41.		
38. Leaves *only* spiny, or reduced to spines 39.		
38. Branches and sometimes also the fruit prickly or spiny 40.		
39. Leaves obovate-oblong, bristly-toothed, in axils of sharp mostly branched spines	*Berberis*,	19
39. Leaves evergreen, oval, margins wavy and with scattered spiny teeth	*Ilex*,	76
40. Vine with tendrils from the stipules; fruit a berry	*Smilax*,	21
40. Not as above 41.		
41. Leaves palmately lobed, often fascicled on the branches	*Ribes*,	58
41. Leaves not palmately lobed 42.		
42. No spines except stunted branches 43.		
42. Stout axillary spines, leaves shining; juice milky	*Toxylon*,	36
43. Pome drupe-like containing 1–5 bony stones; lvs. lobed or wedge-obovate	*Crataegus*,	63
43. Drupe fleshy with bony stone; leaves coarsely or doubly serrate	*Prunus*,	68
43. Pome fleshy, the carpels or cells papery or cartilaginous; leaves ovate, oblong or lanceolate, cut-serrate, toothed or lobed	*Pyrus*,	62
43. Fruit a 2-celled berry; leaves oblong-lanceolate or spatulate-lanceolate	*Lycium*,	105
44. Creeping or trailing evergreen shrubs 45.		
44. Shrubs not creeping 46.		
44. Small or large trees 57.		
45. Leaves small, ovate or oblong; margins revolute; whitened beneath	*Schollera*,	91
45. Leaves small; margins revolute; the lower surface with rigid, rusty bristles	*Chiogenes*,	93
45. Leaves thick, obovate or spatulate, entire, smooth	*Arctostaphylos*,	93

Key to Trees and Shrubs. 125

15. Leaves obovate or oval, obscurely serrate; flowering branches erect......*Gaultheria*. 92
15. Leaves rounded, cordate, bristly, with rusty hairs......*Epigæa*. 92
16. Tendril-bearing vines, with berries in a thyrse......*Vitis*. 80
16. Not tendril-bearing, twining; pods orange-colored, aril-scarlet..*Celastrus*. 77
16. Shrub; leaves evergreen; margins not revolute, green beneath..*Kalmia*. 91
16. Shrub; leaves evergreen; margins strongly revolute, whitened beneath......*Andromeda*. 92
16. Leaves nearly evergreen, ferruginous-scurfy beneath; plant low and much branched......*Chamædaphne*. 92
16. Not as above 17.
   17. Leaves pinnatifid, with many rounded lobes......*Comptonia*. 29
   17. Leaves crenate, obovate, roundish or orbicular, pale beneath..*Betula*. 32
   17. Leaves serrate, toothed or lobed 18.
   17. Leaves entire 53.
18. Leaves roundish, somewhat palmately lobed and cordate; pods purplish......*Opulaster*. 61
18. Leaves sometimes cut-lobed; pome drupe-like; seeds 1-4, bony..*Cratægus*. 63
18. Leaves not palmately lobed 49.
   19. The fruit nut naked, bony, incrusted with white wax......*Myrica*. 29
   19. The fruit berries or berry-like 52.
   19. The fruit a fleshy pome or drupe; trees, often thorny 51.
   19. The fruit dry 50.
50. Leaves long and usually pointed; buds covered by a single scale; fruit a 1-celled pod; seeds numerous, furnished with silky down; shrubs or trees......*Salix*. 30
50. Leaves more or less distinctly 3-ribbed at base, broadly or narrowly oval or ovate; fruit 3-lobed; small shrubs......*Ceanothus*. 79
50. Leaves oval, ovate, wedge-lanceolate or oblong; fruit 5-8 follicles in corymbs or panicles; shrubs, not large......*Spiræa*. 61
50. Leaves broadly-oval or ovate or obovate; stipules oval or oblong-lanceolate; scales of the orbicular or ovate catkins woody; large shrubs......*Alnus*. 32
   51. Leaves ovate or sub-obovate, conspicuously pointed, coarsely or doubly serrate; very veiny; fruit ½-⅔ in. in diameter, the stone smooth......*Prunus*. 68
   51. Leaves ovate, sub-cordate, oblong or lanceolate, toothed or cut-serrate or lobed; the 2-5 carpels or cells papery or cartillaginous......*Pyrus*. 62
   51. Leaves round-ovate, sub-cordate or cuneate, sometimes cut-lobed, large, or smaller and wedge-obovate and oblanceolate; pome with 1-5 bony seeds......*Cratægus*. 63
52. Leaves oblong, mostly entire or slightly toothed; shrub with ash-gray bark and mostly solitary, axillary, peduncled, light-red drupes, with 4-5 bony nutlets......*Iliciodes*. 76
52. Leaves serrulate, with bristly teeth, or minutely ciliolate-serrulate; berry 5-celled, or more or less 10-celled; low shrubs...*Vaccinium*. 93
52. Leaves oval, obovate or wedge-lanceolate, pointed, acute at base, serrate; fruit bright-red, in sessile clusters or solitary......*Ilex*. 76

50. Leaves obovate, or oval, wavy toothed, fruit woody 2-celled......Hamamelis 99

52. Leaves oblong or oblanceolate, acute or acuminate, finely glandular-serrate, tomentose beneath or nearly smooth; fruit red or black..*Aronia.* 62
52. Leaves ovate, ovate-oblong, oblong or broadly-elliptical, rounded or sub-cordate at base, pointed, very sharply or finely serrate; large shrub or tree...............................*Amelanchier.* 62
52. Leaves oval, oblong, or ovate, abruptly pointed, very sharply often doubly serrate with slender teeth, thin; fruit in racemes red, turning to dark crimson; tall shrub, sometimes tree-like ....*Prunus.* 68
52. Leaves oval or oblong-lanceolate, acute or on flowering shoots oblong or obtuse, serrate or finely serrulate; drupe black, 2-3-seeded; shrub low or tall................................*Rhamnus.* 79
   53. Twigs, etc., spicy aromatic; leaves oblong-obovate, pale beneath ....*Benzoin.* 50
   53. The fruit incrusted with white wax; leaves oblong-lanceoate, entire or wavy-toothed toward the apex, shining and resinous dotted...*Myrica.* 29
   53. Shrub with white, soft, very brittle wood, but the fibrous bark remarkably tough; leaves oval-obovate, short petiolate.....*Dirca.* 85
   53. Not as above, but leaves nearly or quite evergreen 54.
54. Leaves very thick, persistent, elliptical or lance-oblong, margins somewhat revolute; capsule 5-celled.................*Rhododendron.* 91
54. Leaves oblong or linear-oblong, persistent, clothed with rusty wool beneath, the margins revolute; slightly fragrant when bruised ....*Ledum.* 91
54. Leaves oblong-obtuse, nearly evergreen, scurfy especially beneath; capsule depressed, 5-celled; low, much-branched.....*Chamædaphne.* 92
54. Not as above 55.
   55. Leaves entire or slightly toothed, oblong; a much-branched shrub with ash-grey bark and mostly solitary, axillary, peduncled, light-red drupes with 4-5 bony nutlets.......*Ilicioides.* 76
   55. Large shrub or tree; branches greenish, streaked with white; leaves clustered at the ends, ovate or oval, long-pointed whitish beneath; the cyme of white flowers and deep blue fruit broad and open.... ....*Cornus.* 89
   55. Shrub with long, recurved, drooping branches; leaves oblong-lanceolate or spatulate-lanceolate, often fascicled, narrowed into a short petiole; flowers and oval orange-red berries fascicled in the axils..*Lycium.* 105
   55. Leaves obovate to oblong-lanceolatd, hairy or bristly, or sometimes downy beneath; flowers large and showy; shrubs mostly smallish; capsule 5-celled, 5-valved.................*Azalea.* 91
   55. Flowers small, fruit dark blue or black, and otherwise not as above 56.
56. Fruit a 10-celled berry containing 10 seed-like nutlets; shrubs much resembling the next, commonly sprinkled with resinous dots..*Gaylussacia.* 93
56. Fruit a 4-5-celled berry or imperfectly 8-10-celled by false partitions, many-seeded; flowers and fruit solitary, clustered or in leafy-bracted racemes................................*Vaccinium.* 93

Key to Trees and Shrubs.

	PAGE
57. Leaves very broad, sharply lobed, palmately veined; bark exfoliating from young trunks and branches annually in plates; branches greenish-white...................*Platanus.*	60
57. Leaves few-lobed, truncate at the appex with a very broad shallow notch..*Liriodendron.*	12
57. Leaves star-shaped with 5-7 long lobes, shining, glandular-serrate; the bark of branchlets usually with corky ridges. *Liquidamber.*	60
57. Leaves not as above, entire 58.	
57. Leaves not as above, denticulate or serrate 61.	
57. Leaves not as above, lobed 60.	
58. Bark spicy-aromatic and very mucilaginous; leaves often lobed..*Sassafras.*	49
58. Not spicy-aromatic; leaves thin, ovate-lanceolate, pointed, 5-10 in. long, 2-4 in. wide, very smooth............................*Asimina.*	12
58. Not spicy-aromatic; leaves thin, oblong, pointed, green and a little pubescent beneath, 5-10 in. long........................*Magnolia.*	11
58. Not spicy-aromatic and not as above 59.	
59. Leaves large, round-cordate, pointed; flowers preceding the leaves..*Cercis.*	69
59. Leaves ovate, pointed, pale beneath, veiny..................*Cornus.*	89
59. Leaves oval or obovate, commonly acuminate, 2-5 in. long......*Nyssa.*	90
59. Leaves thickish, ovate-oblong or elliptic, abruptly acuminate, glaucous beneath, 3-5 in. long; petioles, veins and margins puberulent....................................*Diospyrus.*	95
59. Leaves lanceolate-oblong, thickish, smooth and shining above, downy beneath; fruit an acorn........................*Quercus.*	33
60. Bark spicy-aromatic and very mucilaginous; some leaves entire.*Sassafras.*	49
60. Leaves broad, cordate-ovate, serrate, on young shoots mostly lobed..*Morus.*	36
60. Leaves cut-lobed at the apex, round-ovate or ovate-oblong; pome drupe-like, containing 1-5 bony-seeded stones; large shrub or small tree....................................*Cratægus.*	63
60. Leaves ovate, oblong or lanceolate, acute at base or often rather cordate, cut-serrate or lobed; pome with 2-5 papery or cartilaginous carpels; small tree.....................*Pyrus.*	62
60. Leaves variously lobed; fruit an acorn; large trees..............*Quercus.*	33
61. Leaves finely serrate or denticulate 62.	
61. Leaves sharply, coarsely, doubly, or sinuately serrate 63.	
62. Leaves oblong-lanceolate, pointed, serrulate, on slender petioles; the white flowers and oblong 5-celled pods in long, 1-sided racemes clustered in panicles..................*Oxydendron.*	92
62. Leaves ovate to oblong or broadly elliptical, acute or pointed, very sharply or finely serrate; the flowers and the several or 10-seeded berry-like fruit racemose...................*Amelanchier.*	62
62. Leaves oblong to lanceolate, oval or obovate, finely and sharply serrate, pointed; flowers and 1-seeded fruit in clusters or racemose..*Prunus.*	68
62. Leaves oblong or oblanceolate, finely glandular-serrate, tomentose beneath or nearly smooth; fruit berry-like, 2-5-celled, each 2-seeded..*Aronia.*	62
62. Leaves long, linear-lanceolate or ovate-lanceolate, serrate, denticulate or sub-entire; shrubs and trees, mostly along streams...*Salix.*	30

62. Leaves broad and more or less cordate or ovate, serrulate, serrate, or coarsely toothed; trees with soft, white wood..........*Populus.* 29
62. Leaves coarsely sinuate-toothed; fruit an acorn............*Quercus.* 33
63. Leaves serrate with coarse pointed teeth; nuts in a spiny involucre...*Castanea.* 34
63. Leaves, etc., not as above 64.
64. Leaves broad, cordate-ovate, serrate, usually lobed on young shoots, rough, or smooth and shining above; wood hard, yellowish..*Morus.* 36
64. Leaves broad, cordate, mucronately-serrate; the inner bark very strong; flowers and woody capsules cymous; the peduncle adnate to a large leaf-like bract.................................*Tilia.* 81
64. Leaves rhombic-ovate, acutish at each end, irregularly doubly-serrate or obscurely 9-13-lobed, whitish beneath; bark of young trunks and limbs lacerate-laminate, whitish or ish or reddish tinged ....*Betula.* 32
64. Leaves long, narrow, serrulate or serrate; trees or shrubs growing especially along streams; pod 1-celled, seeds with silky down ....*Salix.* 30
64. Leaves, etc., not as above 65.
65. Leaves strongly straight-veined 68.
65. Veins from the midrib not prominently straight and parallel as above 66.
65. Veins triple at base, somewhat irregularly reticulate..........*Celtis.* 36
66. Trees not at all thorny; leaves broad and more or less cordate or ovate, teeth small and somewhat regular, or sinuate-toothed or crenate...*Populus.* 29
66. Trees not at all thorny; leaves much longer than broad, serrate or serrulate...*Prunus.* 68
66. Trees more or less thorny 67.
67. Thorns conspicuous; leaves round-ovate or more or less wedge-shaped ....*Cratægus.* 63
67. Somewhat thorny by stunted branches; leaves toothed, cut-serrate or lobed..*Pyrus.* 62
68. Twigs and leaves more or less spicy-aromatic; bark somewhat laminate or detaching in thin filmy layers; leaves ovate or oblong-ovate, *cordate or rounded* at base....................*Betula.* 32
68. Not as above 69.
69. Leaves oblong-ovate, taper-pointed, distinctly and often coarsely toothed; nuts sharply 3-sided, in a soft prickly coriaceous involucre; tree with close, smooth, ash-gray bark.*Fagus.* 33
69. Leaves sharply, often doubly serrate; fruit, etc., not as above 70.
70. Large trees; bark deeply furrowed; leaves serrate, unequal at base; the flowers preceding the leaves and winged fruit in fascicles or racemes.......................................*Ulmus.* 35
70. Small trees; bark finely furrowed with narrow longitudinal divisions; fruit resembling that of hops..........................*Ostrya.* 31
70. Small trees; bark smooth, light-gray or ash-colored; a nut at the base of each enlarged scale of the leafy catkin............*Carpinus.* 31

www.ingramcontent.com/pod-product-compliance
Lightning Source LLC
Chambersburg PA
CBHW020113170426
43199CB00009B/512